# ONE TEAM. ONE FIGHT.
# ONE FAMILY

## THE HISTORY OF THE ARMED FORCES
## TAEKWONDO TEAM

## LOUIS E. DAVIS
### WITH MICHAEL BENNETT & RAFAEL MEDINA

*Elite Publications*

An Elite Publications Book Greenville, NC 27858
Tel: 919-618-8075
info@elitepublications.org www.elitepublications.org

One Team. One Fight. One Family.
All Rights Reserved. Published in 2023 by Elite Publications
Copyright text © 2023 by Louis Davis, Michael Bennett and Rafael Medina First Printing: Nov 2023

Library of Congress Control No.: 2023920039
Hardcover ISBN: 9798867157715
Paperback ISBN: 978-1-958037-14-0
Kindle ISBN: 978-1-958037-13-3
Cover & interior design by Krystal Harvey, Tiger Shark, Inc.

Elite Publications does not make any guarantee or other promise regarding the results that may be obtained from using the content of this book. The publisher disclaims any and all liability in the event that any information, logos, names, commentary, analysis, opinions, advice, and/or recommendations contained in this book prove to be inaccurate, incomplete, unreliable, or result in any investment or other losses.

The opinions expressed in this manuscript are solely the opinions of the author and do not represent the opinions or thoughts of the publisher. The author has represented and warranted full ownership and/or legal right to publish all the materials in this book. All attempts have been made to ensure the accuracy of the information presented in this book, but this is not a guarantee.

All rights reserved. This book may not be reproduced, transmitted, or stored in whole or in part by any means, including graphic, electronic, or mechanical without the express written consent of the publisher except in the case of brief quotations embodied in critical articles and reviews.

If you would like to do any of the above, please seek permission by contacting Elite Publications. To order additional copies please visit: https://www.armedforcestaekwondohistory.com/

Printed in the United States Of America

# CONTENTS

Introduction .................................................................................................................. xvii
1. Purpose, Direction, Motivation, Dedication & Determination ................................. 1
2. The Beginning of A Dream ..................................................................................... 5
3. Fort Bragg: Birthplace of The Army Taekwondo ................................................... 9
4. How It All Started ................................................................................................. 11
5. The Navy: It's Not Just a Job, It's An Adventure! ................................................ 19
6. The Air Force: "Aim High, Kick Low!" .................................................................. 43
7. The Army: Be All That You Can Be ..................................................................... 61
8. The Road To CISM ............................................................................................... 89
9. Fort Indiantown Gap: The Rise of The All-Army Team ....................................... 93
10. Into The '90s: The Evolution of The Armed Forces Taekwondo Team ............ 115
11. Gaining Momentum .......................................................................................... 121
12. Michael Bennett And the Fort Hood Taekwondo Team ................................... 131
13. The Paradigm Shift of the 1997 Armed Forces Team ...................................... 135
14. Bennett And Medina Step Up as Coaches ....................................................... 143
15. Interview With Coach Bobby Clayton .............................................................. 149
16. The Return of Bobby Clayton and Rafael Medina ........................................... 161
17. The End of An Era ............................................................................................ 169
18. The Definition of A World Class Athlete .......................................................... 175
19. The Gathering ................................................................................................... 181
In Conclusion ........................................................................................................... 183
Closing Thoughts ..................................................................................................... 199
About The Authors .................................................................................................. 205

*For my wife*

I want to give a VERY special thanks to my wife Pazoo "Lexi" Xiong-Davis. I could not have finished this project without your faith in my abilities as a writer. I know that this project, among other things, has caused me to be forgetful in quite a few areas while working on this project.

In doing so I have found myself putting you and things that are most important to you on the back burner. I therefore dedicate this book, and the hard work put into making this happen, to you. I want you to be proud of me, not just for this project but for simply being the man you love and are building a life with.

# ACKNOWLEDGEMENTS

To Mark Green, I sincerely thank you for your invaluable contributions to the book *One Team. One Fight. One Family. The History of the Armed Forces Taekwondo Team*. You are very well known for your achievements as a soldier and an athlete. Your written work in this project was pivotal in making this book a success, and I am genuinely grateful for your contributions during this time. Your guidance, knowledge, and mentorship encouraged me to step up and step out.

To my two co-authors, Grandmaster/Coach Rafael Medina and Grandmaster Michael Bennett. Coach Medina, thank you most of all for your unwavering confidence in my vision while writing this book. Grandmaster Bennett, thank you for challenging me by introducing me to the Elite Publications team and for your continued guidance, wisdom, and mentorship. I am grateful to you both for joining me as a co-author of this book.

To Grandmaster Sei Hyeok Kim, a special thanks to you for taking the time to meet with me during my visit to Kukkiwon and for sharing your knowledge of Taekwondo with our team through Grandmaster Clayton. I was honored to have stood with you at the 2022 Taekwondo Hall of Fame. Thank you for all you have done for us and our program.

To Grandmaster Robert "Bobby" Clayton. Back in 1997, you believed in my abilities as an athlete as you currently believe in my ability to research and preserve the history of this great Taekwondo Team and the program that came into existence because of it. Because of your faith in me regarding this matter, I can confidently move forward with this book series. Thank you for everything, Sir!

To Master William D. Baldwin. Sir, thank you for opening your home to me and offering me the opportunity to learn firsthand how the Armed Forces Taekwondo Program began. We all owe you a debt of gratitude. Your efforts within US Armed Forces Sports and the USTU gave birth to the US Armed Forces Taekwondo Team. It is my expressed intent that you and the work that you have done will never be forgotten by anyone who wears or once wore the colors of the US Armed Forces Taekwondo Team.

To Master Kevin Williams, aka "Big Dog," you have always been there for me; you stood beside me during my early days with the Fort Hood team when others were ready to step away. Your influence on the coaching staff of this team changed my life. It was truly an honor and a privilege to have stood beside you as a soldier, athlete, and friend. As I prepared to retire from military service, you helped me

ACKNOWLEDGEMENTS

to decide which path was the best to follow. Words cannot express how much your friendship means to me.

To Reginald "Reggie" Perry, you were the sole reason I stayed in the fight and broke barriers not only in my career as a Taekwondo competitor but also as a man of color. I will never forget my time with you in Pickens, South Carolina, and how you reignited the fire that continues to drive me forward. I owe you a debt that would take a lifetime to repay.

To Johnny Birch Jr., whom I've affectionately nicknamed "KO Kid." Over the years, you also stood by me during some challenging times throughout my career as a soldier, athlete, and as a man. Your actions and words reminded me of the importance of standing firm for what I believe in, and you have demonstrated a willingness to stand alone when necessary. Thank you for encouraging me to stand ten toes down!

To Bruce Harris, the OG Coach of the All-Army Taekwondo Team, although our contact with one another has been limited, I would like to thank you for all that you have done to help establish the All-Army Taekwondo Team during your time as head coach. Thank you for all you have done and continue to do to support today's US military Taekwondo athletes.

To Master Kevin L. Jones, my dear friend, brother, and confidant. Throughout this journey to complete the monumental task of writing this book and closing certain chapters in my life, you have stood beside me, offering me words of positive reinforcement, wisdom, and a little bit of humorous insight. It is your sincerity that has earned my trust and my respect. Thank you, brother!

To Mrs. Juliet "Mom" Porter, as your "adopted" son, you once asked me a profound question: "Who are you?" Throughout our journey together and through our many phone conversations, you have challenged me and encouraged me to discover exactly who I am and my true purpose in this life. Thank you for embracing me as a son and for often being that voice of reason whenever my world became chaotic.

To Gregory "Padawan" Sheppard, I will never forget meeting you back in 2003 in South Korea. I never thought that our paths would cross again and that I would be afforded the opportunity to stand beside such a talented athlete, one of the funniest people on the planet (the prank you all played on me back in Aberdeen still makes me laugh), and someone who reminded me that there's nothing wrong with blazing your own path and following your heart. Thank you for reminding me to keep it simple and just be me!

To Master Donovan Rider. Donnie, thank you for affording me the opportunity to visit you at the Charleston Taekwondo dojang. Seeing your unwavering drive and determination to strive for excellence, not only within your students, your staff, and your business model but also within yourself, motivated me to reach for the stars.

To Master James J. Franklin and Master Ricardo Aguilera. James, thank you for encouraging me to return to the Taekwondo world and for exposing a talent that I didn't know that I had as a potential coach. Ricardo, thank you for helping me refine my well-meaning yet crude approach (yes, I said crude) to the sport and coaching. I want to thank both of you specifically because you showed me my limits and my shortcomings, often giving me the old-school tough love that I need from time to time. James, I thank you most of all for the tough love that I've sometimes needed to become a better man. Thank you for always leading by example, an example that I will always strive to follow.

To the Midwest Taekwondo Group, thank you all for offering me a place among you. Even though I am not a school owner, you have treated me with dignity and respect. You will always have my support. *Skol!*

To Mr. Paul J. Boltz and Ms. Claudia Berwager, the two of you were like surrogate parents to us; you loved us despite our shortcomings as athletes and individuals. On behalf of all of us, thank you both for all that you have done for us.

To Grandmaster Gerard Robbins. Sir, I don't have the right words to express my continuing gratitude for offering me a place among the greatest pioneers, practitioners, coaches, teachers, and mentors of our shared martial arts and martial sport. I am truly grateful to you for allowing me to become a part of the ever-evolving history of Taekwondo.

To Mr. Pedro Laboy, thank you for providing me with not only video footage and several lengthy discussions on how you got the ball rolling for the Army but also your willingness to provide me with documentation, which in turn gave me a viable starting point to begin my research. May God bless you and keep you!

To Retired CW4 "Coach" Bongseok Kim. Brother, I could not in good conscience write an acknowledgment section without mentioning you. When you assumed the role as head coach of the All-Army Taekwondo Team, you had some huge shoes to fill. Thank you for giving me a second chance as an athlete and for your no-nonsense approach to our many discussions by phone and in person. I was proud to have had you as a coach during the 2001 season, the 2003 Mid-Atlantic Championships, and when I earned my first national medal at the 2005 USA Taekwondo National Championships. I was even prouder still when you stood with me as I was officially inducted into the 2022 Taekwondo Hall of Fame. Thank you for helping me push past my limits.

To Ron "Hillbilly Gangsta" Berry and James "Jojo" Stagen, I want to thank the both of you for offering me a place among the US Champions Taekwondo Group. Thank you for all that you have done for me and the US Armed Forces Taekwondo Program.

To my daughter, Cecelia Shae Noell Davis. You're the reason I'm still walking a path in this life. God used you as a means to provide me with much-needed light during one of the darkest periods in my life. I hope that when you read this, you will be proud of your father and all that he has done up to this point. *Never forget that you have a father who loves you.*

To my parents, Clifton Lott Sr., Ella Davis-Suggs, and Patricia Lott-Gordon. I don't have the

# ACKNOWLEDGEMENTS

words to thank each of you for all you've taught me and for your input on the man I am in this life. I am proud to be your son.

To my late brother, Clayton Anthony Davis Sr., I hope you're smiling down from heaven and saying to everyone, "That's my brother!" I wish you could still be here to share the success of the book.

To my younger brother, Clifton E.L. Lott Jr. "J.R.," when I retired from the Army, you helped me get on my feet, motivated me to finish my education, and demonstrated the value of critical thinking through your actions! Thank you!

To my sister, Ophelia Lott-Kynnersley, and her loving husband, Neil Kynnersley, the two of you motivated me to make much-needed improvements in my life. You both served as role models for me to get my act together! Thank you!

To my sister, Patrice Ashley Lott, and her future husband, Nai'im, thank you both. Patrice, thanks to our many conversations, you helped me become comfortable with being different and made it work for me.

To my sister, Brittany Stewart, thank you for helping me to get out of my own way. I hope my work will inspire you to reach for the stars and let *nothing* stand in your way.

To Kenneth Garnier. Brother G, thank you for helping me accept the gospel of Jesus Christ on my own terms and for helping me take my first steps into a larger world.

To my dear friend and fellow veteran Rory Lucas. Man, I want to thank you for supporting my efforts to get this done and do so on my own terms.

To the Elite Publications Team, thank you for offering to work with me to make this dream a reality. May God bless you and keep you.

Lastly, all honor, glory, and praise go to God, my Savior, Jesus Christ. Through God, I've been blessed to have served in the US Armed Forces as a soldier, athlete, veteran, and now an author.

*Louis E Davis*

# FOREWORD: THE TRUE MEANING BEHIND "ONE TEAM, ONE FIGHT!"

As individuals, we believe that anything can be achieved. However, not everything can be achieved alone. Family, friends, teammates, and coaches are all sources of help and guidance. Without these sources, it would be impossible to win every fight and overcome every struggle. Instead of being confined to Taekwondo or even athletics in general, "one team, one fight" should be applied to ordinary, everyday situations.

It is a saying that should be especially remembered during the worst of moments, when one's team is needed most. On the other hand, the corresponding symbol more lucidly represents martial arts. Together, the symbol and motto most strongly uphold the antiquated ideas of teamwork and unity, which are both important aspects of life.

There are moments during human existence that require the assistance of another. Maybe a pass needs to be thrown to an available teammate, or perhaps advice is requested on various subjects. Whatever the case, there is one similarity: teamwork. For example, it takes a team of nine to win a baseball game, just as it takes a team of qualified doctors to perform open-heart surgery. Likewise, it takes an entire pride of lions to capture and kill their prey, and a whole team of astronauts to explore the worlds outside of Earth's atmosphere. Similarly, a team of instructors, friends, and family is necessary in Taekwondo. Although Taekwondo is not exactly a team sport, more than one person can be actively involved in a single sparring match or a full tournament. An experienced teacher can steadily improve any student's technique and form. Emotions which run high, especially during tournament time, can be nurtured by friends or family. Fellow classmates can also provide their own words of encouragement.

Despite an array of differences, the spirit of Taekwondo is always able to join diverse people into one common bond. As a group, they are able work towards personal goals and hopefully accomplish them, e.g. victory, physical/emotional improvement... Believe it or not, more can be attained through teamwork than individually. However, a team consists of individuals, each with his/her own strengths and weaknesses. These characteristics are augmented during a tournament, where competitors from across the globe compete for gold and glory.

# FOREWORD

The symbol more fully represents this competitive aspect of Taekwondo. It shows that the meaning of victory/defeat lies between two fighters. Even though the competitors may not know each other and come from contrasting areas of the world, they still meet for the sake of competition. Despite the fierce violence of a sparring match, tournaments are approached in a controlled, peaceful manner. Also, the tenets of Taekwondo (especially self-control) are greatly exhibited; sportsmanship is a must-have. Nonetheless, it took a team to get that student into the ring in the first place.

It takes a team to make a champion: one team, one fight. It takes a village to raise a child. Without aid or support, everyone would be homeless. "One team, one fight" should spread from Taekwondo and be considered a part of our daily routine. Although its counterpart—the symbol—is more dedicated to Taekwondo, it should be equally respected. After all, competition is what fuels all sports. In conclusion, the impossible is foolish to attempt alone but with the support of a team the impossible can be accomplished.

**Carolyn Carney, 1st Dan**
July 21, 2003
Sport Taekwondo Center

# PREFACE

The military athlete is a different breed of athlete; regardless of age, rank or branch of service, men and women who serve go through their own private hell to earn not only the privilege of wearing their branch's colors, but also the privilege of competing, representing said branch of service and this great nation of ours.

It's through this little-known process that they become more than "just" Soldiers, Sailors, Airmen or Marines; in their pursuit of athletic excellence, they become something larger than life; they become unofficial ambassadors of the United States of America promoting peace by upholding a simple yet profound message: "Friendship Through Sport." It's a privilege that money simply *cannot* buy.

Their respective battlegrounds shift from the field of warfare to the field of athletic competition. What begins as fierce rivalry quickly becomes loyal friendship, friendships that can and often last a lifetime. It is this friendship that has the power to end wars, to unite people in a way that politicians can only dream of.

Before we begin, I want to thank Colonel Mark Green for encouraging me to "Step up and Step Out" and get this project done. I also want to thank GM Reginald Perry for sparking my curiosity which led to me writing this book, to GM Rafael Medina, GM Michael Bennett, GM Bobby Clayton, Master Chief William Baldwin, Sergeant Major Bruce Harris, Sergeant Major Edward Fourquet, Lieutenant Colonel Punarrin Koy, Colonel Howard Clayton, Chief Warrant Officer Bongseok Kim, Master Kevin Williams, Master Johnny Birch Jr., Master Todd Angel, Master Javier Martinez, Master Darrell Rydholm, GM George Bell, Petra Kaui, Nicolau Andradae, Yelena Pisarenko, Jada Monroe, Mr. Paul Boltz (RIP), Ms. Claudia Berwager, GM Ron Berry, GM Larry Spears, GM Luis De La Rosa, GM Sterling Chase, GM Patrice Remarck, GM John Holloway, GM Gerard Robbins, GM Dwayne Harris, GM Jay Dunston, GM Barry Partridge, GM Salim J. Oden, GM George Bell, Professor Steven Capener, GM Sei Hyeok Kim, Master Jojo Stagen, Master Joe Lupo, Master Trinity Trongg Osborn, GM Peter Bardatsos, GM Jeff Pinaroc, GM Ricardo Aguilera and GM James Franklin, Sergeant First Class Jonathan Fennell, Sergeant Gregory Sheppard, Sergeant Donovan Ryder, Master Owen Brown, Master Carlos Rentas, Master Kevin Louis Jones, Master Salim Oden Jr., Master Jody Gibson, Master Brad Carter, Master Edward Givans, Master Eric Laurin, GM Jody Gibson and Mr. Jordan "JJ" Perry.

Lastly, I want to thank the US Champions group, USA Masters Team, and the US Military Taekwondo Foundation for offering me a place among you. I know that there's quite a few more names I owe a debt of gratitude to. Please understand that this list would probably fill a few pages. In the spirit of keeping it simple: thank you, each and every one of you!

# NAMES ON A FORGOTTEN WALL

William D. Baldwin, Daryl T. Kubotsu, Craig Lightfoot, George Nobles, Rafael Medina, Pedro Laboy, Mark Green, Steve Brown, Phillip Cota, Paul J. Boltz, Claudia Berwager, Bruce Harris, Bobby Clayton, Michael Bennett, Bongseok Kim, Hyun Suk Lee, David Bartlett, Kevin Williams, Jonathan Fennell, Patrice Remarck, Punarrin Koy, Terrence Jennings

To many, these are simple names of some long-forgotten people who were relevant at some point in time; to others they are nothing more than individuals who, in some capacity, served the US Armed Forces. To those of us who intimately knew these names, benefited from their hard work, dedication and support, their names mean so much more.

Without these brave individuals, there would be no All-Army, All-Navy, All-Air Force, All-Marine and no Armed Forces Taekwondo team. This book, which contains a collection of both pictures and one-on-one interviews, serves to honor their memory by remembering their efforts to make the "Program" a reality.

As mentioned by Rafael Medina, "When you pass away, all of your history, your accomplishments, all of the things that give you and others who knew you a sense of pride will leave with you..."
This book is my personal tribute to those who built the road of the Armed Forces Taekwondo Team that I once traveled upon. To me, each of these names is owed an overdue expression of gratitude for a program that spanned well over 20 years.

A program that produced a myriad of national and international champions, world-class athletes, coaches, teachers, mentors and brotherhood that goes beyond the mat, beyond the dobok, beyond the belt ranks and beyond the uniform. One team, regardless of military rank, branch of service and military occupational specialty, stands united on the field of competition, one common fight; be it on the competition mats or in the theatre of war we stand shoulder to shoulder. One family due to those strong and nearly unbreakable bonds forged through military service and the pursuit of athletic excellence.

**THIS is also the true meaning behind "One Team, One Fight, One Family!"**

# INTRODUCTION

The US Armed Forces in general and the US Army in particular, are known worldwide for being an all- volunteer military force, for its war-fighting ability, for accepting all who serve regardless of background to include race/ethnicity, creed, color, religion, gender, and sexual orientation. For its structure and discipline and the list goes on.

However, within this glorious organization there exists a group of unsung heroes who without the help of the internet, and interested individuals such as myself, would fade into obscurity. Military athletes in general, Taekwondo in particular, are among these heroes. The Department of Defense aka "DOD" (as I've come to learn) doesn't always keep an accurate record of its pioneering athletes, coaches, and competitors past and present, therefore it is imperative that this book exists.

It is my expressed hope that this book will offer accurate knowledge of how the team came into being, who were the early pioneers of what many refer to as "The Program." This will be shared through firsthand accounts from individuals who were there during its development, through its evolution, prior to during and two different times of war (Operation Desert Shield/Desert Storm and prior to 9/11: War On Terrorism), through the establishment of the Army's WCAP (World Class Athlete Program), the creation of Korea's 2ID Taekwondo Team, the creation of the unofficial Fort Campbell Taekwondo Team, the creation of the Fort Hood Taekwondo team, and the Fort Lewis Taekwondo team.

This book will attempt to offer insight into exactly what a Military World Class Athlete is and the burden that they carry, and the struggle the team has had with the national governing body (the former United States Taekwondo Union or USTU, and USA Taekwondo or USAT). Finally, the creation of the US Military Taekwondo Association and the US Champions Facebook Group. Both organizations serve to maintain our strong bonds as athletes and as veterans.

**Try to imagine...**

Try to imagine October 2015, Hinesville, Georgia...the event is the Liberty Taekwondo Championships.

To many it's just another in a long line of Taekwondo competitions within the state of Georgia but to a small group of individuals, this event is something much more.

This event also serves as the Armed Forces Taekwondo team reunion. This was our second reunion. As a resident historian on what many of us refer to as simply "The Program," I've taken it upon myself to ensure that these legends and the sacrifices that each of these brave men

# INTRODUCTION

and women have made for the program are remembered.

There are many people to include, even our very own service members who are completely unaware of the fact that all four branches of the US military have intramural sports teams such as this one; some of these sports are team sports and others considered are individual sports.

Try to imagine competing in any of these sports which have previously led these service members to the Olympic games and a very select few to the coveted Olympic Gold Medal, and fewer still into a professional career in their respective sports. (Boxing's Ray Mercer and Riddick Bowe are two very good examples of this.)

As I've stated earlier, the US military athlete is a different breed of athlete, for they must at all times wear two hats: the chief of these two hats being the defense of their nation and its way of life, the other, maintaining one's competitive edge in their respective sport.

It is widely known that the US Armed Forces is an all-volunteer force, that's right, an all-*volunteer* force (of course they pay us lol), but every service member who chooses to compete in an intramural sport as a representative of their respective branch of service, regardless of job classification, rank and file, has to endure their own personal and professional hardship, to include enduring their own private hell just to prepare for the team trials in their respective sport.

Try to imagine that hell being described as the unit's mission (which has priority over all else), to include that unit's leadership (especially if you are a junior grade enlisted or worse, a low-ranking officer); then there's the jealousy of one's leadership whose opinion is, and I quote, "You joined the military to be a (enter military job here), not play 'xyz' sport." For them your only purpose in life during your enlistment is to soldier and get promoted, no more, no less.

Try to imagine wanting to be something more, try to imagine wanting to do something that stands out during peacetime. This is the struggle all aspiring military athletes have had and continue to have.

# CHAPTER 1
# WHY AM I DOING THIS?
## PURPOSE, DIRECTION, MOTIVATION, DEDICATION & DETERMINATION

2016-Armed Forces Taekwondo Reunion, Hinesville, GA

So many of you are probably wondering, why am I taking the time to write about this? That answer is as simple as it is complicated, but my primary reason is to preserve the Armed Forces Taekwondo program's rich and little-known history. The other reason is to honor those who laid the foundation for each of the teams representing the four main branches of the US military in the sport of Taekwondo, and to encourage the future generations who will carry our torch.

Many of you are already thinking, "Who cares about the history of the Armed Forces Taekwondo Team?" Well...YOU SHOULD! THOSE OF YOU WHO ARE WEARING THE UNIFORM!

HELL, EVEN YOU, AMERICAN CITIZENS, YOU MOST OF ALL SHOULD CARE! Why, you ask? Because we as military athletes represent you!

Please understand, as I tell this story I will attempt to do so not only through my own personal experiences but through the eyes and firsthand experiences of my fellow teammates, coaches both current and former, through the eyes of our beloved sport directors Mr. Paul Boltz and his successor Ms. Claudia Berwager, and through the eyes of Bruce Harris, the All-Army team's second official head coach; through the eyes of everyone that I've been able to find that were actually there at "Ground Zero" when the program was born.

Then there are the firsthand accounts of top veteran players such as Reginald Perry, Bobby Clayton, Carlos Rentas, Edward Givans, Brad Carter and Bongseok Kim, through the eyes of the first soldiers to represent the Army in the sport and who are considered the pioneers of the All-Army Taekwondo Team: Pedro Laboy, Rafael Medina, and Mark Green.

Many who read my writings will disagree with me on this, but each of these individuals, to include myself and countless others, has helped to build a program which has endured the ever-looming threat of being cut due to the budgeting issues with MWR (Morale, Welfare and Recreation), and overall low numbers and politics within Taekwondo's governing body USTU, then later USAT. Most of those low numbers of applicants to our trial camps stem from two wars, namely Operation Desert Shield/Desert Storm and the war on terrorism, and currently the Covid pandemic.

It is my expressed intent to enable all who read this to see things through the eyes of the athletes and the support staff responsible for laying the foundation of our beloved Taekwondo program, and more importantly than that, to honor them. This is my Love Letter to the Armed Forces Taekwondo program.

My inspiration to uncover the events leading up to the creation of "The Program" came at a time when I had lost sight of what it means to be a Soldier-athlete. Of course, there were people within the ranks eager to "remind me" that I am a soldier first and an athlete second. As a result, I was given 45 days to "reacquaint myself" with this concept.

After finishing the last day of my 45-day "reacquaintance," the result of a field grade "Article 15" (don't ask), my first order of business was to get back what my unit, the 12th Chemical Company, attempted to take from me: All-Army Sports, and my dream of qualifying for the assignment to the Army World Class Athlete Program, the next step to pursuing Olympic Gold.

I was eager to take a well-earned 30 days of leave to go to Pickens, South Carolina to visit one of the All-Army team's legendary fighters, Mr. Reginald Perry; it was Reggio along with Mr. Kenneth Carter Sr. and Andrew Blocker who were instrumental in helping me refocus on the task ahead, which was to retake my place on the team.

The first night I spent at Reggie's trailer out in the rural hills of Pickens, and I vividly remember when I first set foot in this small dojo connected to the trailer; it felt like I had stepped back in time. Hanging on one side of the wall were all of his newspaper articles, the medals he had won as a competitor, and all of his team photos.

I remember listening intently to how passionately Reggie spoke about the team, how he felt about representing the Army. All these things left me with a burning question: "When, where and most of all how did this all start?! Whose idea was it?"

I found myself asking many former athletes from that generation, to include Reggie himself, this same question, and nearly all of them seemed to unanimously mention three men: Rafael Medina, Pedro Laboy and Mark Green.

It was Reggie's passion about All-Army Taekwondo that renewed my motivation to retake my place on the team; it also led me to start researching the team's history with the help of Coach Medina.

# CHAPTER 2
# THE BEGINNING OF A DREAM

Louis Davis (Blue) competing in the 2000 Bavarian Championships

What few people outside of our "family" know is, where was the All-Army Taekwondo Team created and more importantly, when was it created?

In early 2007, GM Rafael Medina aka "Coach Medina" hosted a website containing clues as to the origins of the Army and the Armed Forces program. In his section, which was dedicated to the Armed Forces, is contained the following in his own words:

"Many sports will keep soldiers in top physical condition if the soldier maintains the sport as a serious business, but to be one of the best takes time and a lot of discipline. The task of being a soldier and an athlete is not an easy task. You must get up early in the morning and get ready for PT, maintain your job skills, and perform any additional duties or tasks that lie ahead. Soldiers are soldiers twenty-four hours a day for seven days a week. However, becoming a soldier and an athlete is twice as hard; sacrifices in both your personal and family time must be made in order for you to get into top physical condition.

Lots of soldiers studied Taekwondo before us, but as far as I know, they never represented the Army as a team. It was both myself and my close friend, Pedro Laboy, who took the first steps to officially represent the Army. At that time, the Navy and the Air Force had established teams before we did.

When I reflect back to 1984 and all the things we did to gain support from our chain of command, it began with us training two hours prior PT formation (6:00 a.m.) in order to prepare for both the North Carolina State and the US National Taekwondo Championships.

At that time there were many obstacles that we had to overcome, first of all there was absolutely no support from our company commander, then there was the unit's mission, details, field training exercises, weapons qualification ranges, physical fitness tests and so on, but we were determined to see this through.

All we had at that time was a shared vision, a dream and in order to accomplish that dream, we needed to first prove to our leadership at Fort Bragg (and later to the Army) that by demonstrating the highest quality of soldiering and athleticism we could effectively represent them well in this sport.

Thanks to their success in 1984, the following year at Fort Bragg, North Carolina, the Army Taekwondo Team was unofficially formed. Reading and hearing about what these pioneering soldiers went through in order to have a team representing the Army in this sport gave me and continues to give me a sense of pride; I am honored to be a part of this rich and nearly forgotten history which continues to be written even to this day.

Unfortunately, the generations that rose after the last decade and a half have seemingly forgotten about these people who've laid the foundation for the program that they've inherited, and the heavy burden of keeping the program alive by introducing it to new generations of future Armed Forces athletes. Sadly, as I write this it is doubtful that any of them would care.

During my five years' stay in Germany, I used the things that the Fort Bragg Taekwondo team endured to enable them to compete as my guide into training myself in order to maintain my own competitive edge. My reasoning was 'If the Fort Bragg guys trained two hours before PT every morning, then I shall do the same. I will follow the same path that they once followed.'

For me that path consisted of a 2 – 3-mile run followed by short distance sprints and ending with 3 rounds of 3 minutes of skipping rope with a 30 second rest in between each round. Finishing by 0600 with more than enough time to change into my PT uniform and head to first formation.

The second phase of that path consisted of sacrificing my lunch time to get into the Harvey Barracks gym to run on the treadmill for 6 minutes, 3 rounds of 3 minutes of skipping rope with a 30 second rest in between each round and (when it was available) kicking the heavy-bag or the basic line drills that I learned from both Coach Bennett and Coach Angel while stationed at Fort Hood, or the line drills I learned from the All-Army Camp at Fort Indiantown Gap.

The third phase was evening training, which initially began with training at a local Taekwondo school in Kitzingen, then later in Würzburg under Peter Müller at his school known as TGW Würzburg. Peter's group was a tough as nails group EACH specializing in either Poomse, self-defense or sparring.

Their group led me to my first regional championship, which I won quite convincingly. The competition was held in a town called Bad Windsheim. This event qualified me to compete in the Bavarian Championships (In German, Bayern Meisterschaft). To my knowledge, I'm the only US soldier to have not only competed but win this competition consecutively (1998 – 2000).

But that is a story for a different time..."

# CHAPTER 3
# FORT BRAGG:
## BIRTHPLACE OF THE ARMY TAEKWONDO TEAM

The famous black T-shirt with gold letters (1986)

During my time in Germany, I found myself constantly speaking with Rafael by phone for ways to train myself so that I could be in top condition when I returned to All-Army Team Trials in 1999. I was determined to be the best in my division. It was here that Rafael told me how the team was formed and that he, his friend Pedro Laboy and Mark Green pioneered what would later become the All-Army Taekwondo Team.

He spoke eloquently about how the three of them had to train with *no* support from their respective units, and that their units' respective mission had priority over all else. He went on to mention how they had trained two hours before first formation which is usually at 0630 (meaning you had to be standing in formation at that time). Armed with this knowledge I too began to train at 0400 and went to PT afterwards. No matter which of them I spoke to, they all said the same thing. Try to imagine soldiering *and* trying to train for an event like the State Championships or US National Championships; I'll add one better, try to imagine having to go into the "field" for 30-45 days for training exercises and upon returning have maybe a week to two weeks to prepare. Well, *that* my friend, is what these three men had to contend with.

After talking with Pedro, I learned that he and Rafael were stationed together in the 82nd Airborne Division and that they were both from the same place: Puerto Rico. They spoke a common language, Español (Spanish). Pedro himself was a member of Puerto Rico's National Taekwondo team at that time and Rafael (as I later learned) was a Black Belt in Kyokushin Karate and a Red Belt in ITF Taekwondo, while Mark Green, who joined them a short time later, was also a Black Belt in Taekwondo.

Both Rafael and Pedro began teaching martial arts out of their respective houses to the local children living in the base housing at that time. What isn't very clear was when they decided to compete, and compete representing both the Army and Fort Bragg, and more importantly *why* they chose this path.

This quest came with its own unique set of challenges, chief of which being, how would they be able to balance their military duties to include: PT, their duties under their respective MOS (Military Occupational Specialty, aka their job), any field training exercises, weapons qualification and any "additional" tasks that their respective leadership may have for them; then there are their families (all three men were married).

Nevertheless, they made it happen. With Pedro calling the plays, i.e., dictating the training, these three brave 82nd Airborne paratroopers made use of the time in between and prior to their military responsibilities. Their hard work paid off; the three of them entered the North Carolina State Taekwondo Championships and *all three* of them won in their respective weight divisions, qualifying them to compete in the USTU National Championships later that year in Dayton, Ohio.

Their success also gained the attention of the local newspaper, and of course not only did their respective chains of command within their units begin to support them, but they gained the much-needed support of the 82nd Airborne Division and additional support from the department of Morale, Welfare and Recreation (aka MWR).

This came in the form of time allotted to them by their chains of command to train and hone their skills in preparation for the national championships. Initially they trained on their own before PT from 0400 to 0600. Now with the blessing of the 82nd Airborne Division, they were allowed to train from 0400-0600, then from 1300-1600, and lastly from 1800-2000.

They were able to attend their first national championships using funding left over from Fort Bragg's boxing program, which covered their travel to and from the event and their food and lodging. Their only uniform was a black T-shirt. It was this first showing at the USTU National Championships that would lay the groundwork for what would later become the All-Army Taekwondo Team.

# CHAPTER 4
# HOW IT ALL STARTED

While reaching for clues to answer the myriad of questions that I had regarding the history of the Armed Forces Taekwondo Team and the teams representing each branch of the US military, GM Medina reminded me of a letter written by Air Force Captain Daryl T. Kubotsu.

The letter itself identifies four associations representing the four branches of our nation's armed forces: Chief Petty Officer William D. Baldwin, President of the US Navy Taekwondo Association; Sergeant Pedro Laboy, President of the US Army Taekwondo Association; Major Craig Lightfoot, President of the US Air Force Taekwondo Association; and Captain George Nobles, President of the US Marine Corps Taekwondo Association.

The very existence of these respective associations fell under a larger umbrella known as the Armed Forces Taekwondo Association. An article written by Captain Kubotsu and his wife Janice provided some much-needed insight on who, what, where, how, and why the Armed Forces Taekwondo Team was created.

What the article states is the following: in 1983 the president of the AAU (Amateur Athletic Union), Dr. Dong Ja Yang, saw a need to establish a chapter within the AAU specifically for the US military similar in nature to what the South Korean Armed Forces had established at that time.

Both Air Force Captain Daryl T. Kubotsu and Navy Chief Petty Officer William D. Baldwin were chosen by Dr. Yang to develop a Taekwondo association for their respective branches of service. A letter written by Dong Ja Yang was given to both men authorizing them to begin the creation of their Taekwondo teams.

By 1984, the Navy and Air Force Taekwondo associations were established, fully operational, and in the process of planning their own separate championships, which would be similar to the state-level championships already established by what would later be known as the US Taekwondo Union.

Both men maintained close communication with one another, working together to prepare their athletes and ensure their respective team's readiness to deploy to the theatre of athletic competition both nationally and internationally.

On November 22, 1985, at approximately 1800 hours (6:00 p.m.) in Las Vegas, Nevada, the very first all-military Taekwondo championships were held. This event was hosted by

Master Dong Sup Kim at his dojang. GM Kim not only hosted the event, but he also provided the medals for this event.

Initially, the winners of each weight class would compete against members of the Kuwait National Guard's Taekwondo team the following day. Unfortunately, the competition with the Kuwaiti National Guard would not take place due to the Kuwaiti team not being able to make the journey.

This military competition later became known as the Southwest Championships thanks in part to Dong Sup Kim, California State Taekwondo Association President GM Chang Yong Kim, and the US Armed Forces.

The Armed Forces team competed against state teams from California and Nevada respectively, with the Armed Forces winning quite a few of these matches, and the US Navy capturing the top slot at this competition.

# The NATIONAL AAU TAEKWONDO COMMITTEE, INC., OF THE UNITED STATES

- National Sports Governing Body for Taekwondo
- A Member of the United States Olympic Committee
- Established the Service Agreement with the Amateur Athletic Union of the United States of America

Prof. Dong Ja Yang, President
Professor in P.E., Health, and Recreation
The National AAU Taekwondo Committee, Inc.
10 Treadway Court
Brookeville, MD 20729
U.S.A.
Office (202) 636-7226
Residence (301) 774-0359

June 30, 1981

Mr. William D. Baldwin
8919 A London St.
Norfolk, Va. 23503

Dear Mr. Baldwin,

This is to certify that you have been appointed to serve as a coordinator in an effort to organize an effective Taekwondo program for the United States Navy and its community.

I trust that you will accept the appointment and conduct appropriate endeavours accordingly.

Thank you,

Sincerely,

Dong Ja Yang
President

U.S. Sole Representative Organization to the World Taekwondo Federation and Kukkiwon

Letter from AAU President Dr. Dong Ja Yang to Chief Baldwin

## 1ST U.S. MILITARY TAE KWON DO CHAMPIONSHIPS
### A Martial Art Military Tradition Has Begun!
by Daryl and Janice Kubotsu

Senior Airman Anthony B. Herod (Air Force) throws a jumping roundhouse to his opponent.

Tae Kwon Do is gaining popularity in the United States Military for a number of reasons. The martial art has long been a useful conditioning and discipline tool, as well as an effective combative art. During the Vietnam Conflict, special advisors of the R.O.K. Army trained both American and South Vietnamese Special Forces in Tae Kwon Do. Today it is not uncommon to see U.S. Military personnel assigned near Panmunjom training in Tae Kwon Do next to their South Korean counterparts. Other U.S. Army Forces personnel stationed in the "Land of the Morning Calm" have likewise had the privilege to study Tae Kwon Do. The word "Tae Kwon Do" is becoming a common term on U.S. military installations as directors of recreation centers attempt to select good programs that a military person and his or her family may be able to enjoy.

In 1983, Dr. Dong Ja Yang, then President of the National AAU Tae Kwon Do Union (presently the United States Tae Kwon Do Union), saw a distinct need to establish military chapters similar to the role model already established in Korea. Air Force Captain Daryl T. Kubotsu and Naval Chief Petty Officer William Douglas Baldwin were chosen to develop an Air Force and a Naval Tae Kwon Do Association. By 1984, both chapters became operational, planning separate championships similar to state level competitions already established within the USTU framework. In 1985, Chung Sik Choi of the United States Army earned a berth on the United States National Tae Kwon Do Team. Other military Tae Kwon Do athletes earned invitations to participate at national events sanctioned by the USTU such as the U.S. Olympic Sports Festival.

Capt. Kubotsu and CPO Baldwin stayed in close communication, each feeling a need for additional competition to prepare their Tae Kwon Do athletes for both national and international competition. At 6:00 P.M., on the 22nd of November, 1985, their dream became a reality as the 1st U.S. Military Tae Kwon Do Championships were opened at the dojang of Dong Sup Kim of Las Vegas, Nevada. Each weight division champion was to face a respective weight champion of the Kuwait National Guard Tae Kwon Do Team the following day at the 1st Western U.S. Tae Kwon Do Championships. However during the height of the exciting competitions, the Kuwait National Guard team captain called to notify the U.S. military members that they would not be able to make it for the friendship match. Through the efforts of Dong Sup Kim and Chang Yong Kim of California, the military team was given additional matches with the Nevada State and California State Tae Kwon Do Teams. The evening ended with the following team results: 1. United States Navy, 2. United States Air Force, 3. United States Marine Corps, and 4. United States Army.

Work is already in progress for the 2nd U.S. Military Tae Kwon Do Championships to be held in San Francisco or Virginia Beach in fall, 1986. This year's military champion will be given additional training in hopes of a friendship match with the South Korean National Military Team in 1987.

Article about the 1st US Military Championships featured in the *Taekwondo Times Magazine*, September 1986, written by Daryl and Janice Kubotsu

# ARMED FORCES TAEKWONDO ASSOCIATION

22 August 1986

Mr. William D. Baldwin, USN TKD Assn
Mr. Pedro Laboy, USA TKD Assn
Capt. George Nobles, USMC TKD Assn
Maj. Craig Lightfoot, USAF TKD Assn.

Dear sirs;

Greetings. We sincerely hoped that you enjoyed the article on the 1st U.S. Military Taekwondo Championships in the Sept. 86' issue of Taekwondo Times. Well, it is already that time of the year for our second championships.

During the nationals, Masters Jerome Reitenbach and James Wilson expressed an interest to hold the second military championships in their geographic areas (San Francisco Presidio and San Diego respectively). After writing follow up letters, no response has been received. My own Grandmaster, Dong Sup Kim of Las Vegas, has offered to give us one ring to use during the 2nd Western Regional Taekwondo Championships. Our students from Hawaii had already planned to attend this event, and so we are in favor of this recommendation. However, I would like some feedback from the rest of you as soon as possible.

I did talk with George Nobles about a month ago concerning a possible championship in Washington, D.C. This would be alright as long as we could get travel funding and accomodations on base (i.e. Bolling, Andrews, etc.). Please call me as soon as possible at work: 449-9907/9908 home: 808-423-1592. We need to make a decision quickly, inorder to get the word out to the field.

Sincerely,

Daryl T. Kubotsu, Capt, USAF

Daryl & Janice Kubotsu
7110-A Ohana Nui Circle
Honolulu, Hawaii 96818

P.S. - I did get a nice letter from Kukkiwon. They are very interested in Friendship Matches between national military associations and possibly another CISM Taekwondo Championships.

Letter from Daryl Kubotsu addressing the Armed Forces Taekwondo Association

AFZA-PA-M

SUBJECT: Army Karate Team

THRU: Commander
US Army Forces Command
ATTN: AFPR-PSM
Fort Gillem, Georgia 30050

TO: HQDA (DACF-LS)
Alexandria, Virginia 22331-0512

The attached applicants' applications (Laboy-Perez, Pedro F., E-4, Green, Mark, E-4 and Medina, Rafael A., SGT) are forwarded for your consideration for selection to the Army Karate Team.

FOR THE COMMANDER:

3 Atchs

FRANK C. RAUCH
Colonel, FA
DPCA

TDY Orders for the 1987 Armed Forces trial camp at Little Creek, Virginia, provided by Rafael Medina and Pedro Laboy

# The First USTU Military Championships—A History

Taekwondo has long been considered an effective training discipline in the military. During the Vietnam War, special Korean advisors trained American and Vietnamese Special Forces in Taekwondo. It continues to be a popular activity for troops stationed in Korea. Americans stationed at Panmunjom are often seen in Taekwondo training alongside their South Korean counterparts.

By 1976, Taekwondo practice was sufficiently well established in the military for the Council of International Military Sports (CIMS) to adopt it as its twenty-third official sport. In 1980, the first CIMS Taekwondo Championships were held in Seoul, Korea. One hundred forty-seven contestants from fifteen nations took part.

In 1983, the NAAUTU moved to organize and encourage the development of Taekwondo in the U.S. military by establishing official amateur Taekwondo associations in the armed forces. Then-president Dr. Dong Ja Yang appointed Air Force Captain Daryl T. Kubotsu and Navy Chief Petty Officer William D. Baldwin to head up these efforts.

Both men were re-appointed in 1984 by incoming USTU president Master Moo Yong Lee, and it is largely through their efforts that the Air Force and Navy were able to hold their own championships in 1984, parallel to the state Taekwondo association championships. Winners were able to take part in the 11th USTU National Championships held in June of that year in Hartford, Connecticut. The four Air Force and three Navy competitors were joined by a large contingent of Army athletes from Fort Bevins, Massachusetts and Fort Bragg, North Carolina. Chung Sik Choi of Fort Bevins earned a berth on the United States national team. Other military athletes were invited to participate in the National Sports Festival in Baton Rouge, Louisiana, the following month.

Observers at the National Championships were enthusiastic about the performance of the military athletes, but felt their overall success was hampered by a lack of tournament experience. This led to plans for the First USTU Military Championships, an interservice competition that would prepare military athletes for future national and international competition. The Air Force was extremely lucky to have Mr. Dae Sung Lee, a member of the U.S. National Team, volunteer to serve as their coach.

Baldwin and Kubotsu stayed in close communication after the nationals to formulate plans for the interservice competition. Since the military had no USTU-certified referees, it was decided to schedule the Military Championships in conjunction with a civilian USTU tournament. Master Dong Sup Kim had already planned a U.S. Taekwondo Championship in Nevada that was to feature a match with a team representing the Kuwait National Guard. Arrangements were made to hold the U.S. military championships in Nevada, one day prior to Mr. Kim's tournament. This enabled the military personnel to participate in the USTU Referee Seminar.

USTU Vice President Jerome Reitenbach presided over the tournament. First place winners (see below) took their places on the U.S. Military Taekwondo Team, to compete against the Nevada State and Kuwait National Guard teams on the following day.

A post-tournament debriefing was held to assess the event and make plans for the future. As most participants in the event had been from the western U.S., it was decided to schedule another competition on the east coast, to take place in February, 1986, in Norfolk, Virginia. Further, the decision was made to look into holding a training camp for military athletes at the U.S. Olympic Training Center in Colorado Springs. The U.S. Military Taekwondo Committee adopted a policy to encourage dojang development at military installations, as well as encouraging the development of military athletes in civilian dojangs. The committee also hopes to sponsor a CIMS Taekwondo Championship in the United States at a future date.

---

**1st USTU Military Championships: First Place Winners**

Fin— none
Fly— J. Gonzales (Air Force)
Bantam— E. Harris (Air Force)
Feather— R. Spears (Marine Corps)
Light— L. Rodda (Navy)
Welter— M. Hampton (Navy)
Middle— A. Herod (Air Force)
Heavy— J. Hufford (Air Force)

Women— Airman 1st Class Robin Reck (Air Force)

1ST USTU Armed Forces Invitational

# The 2nd USTU Military Championships
## Las Vegas, Nevada, CA • November 15, 1986

### by Craig Lightfoot

The U.S. Air Force and U.S. Navy Taekwondo Associations renewed their duel for the coveted team title with Navy coming away with the title for the second year in a row. Supported again by Master Instructors from the western United States and the World Taekwondo Union, the event was a huge success. Grandmaster Nam Suk Lee of the World Taekwondo Chang Moo Kwan presided as the honorary President.

According to Masters Baldwin and Kubotsu this is a very important year for the military athlete. The 2nd CISM Taekwondo Championships will be held this year in Seoul, Korea. The U.S. Military Team will be competing against more experienced teams from throughout the world. The eyes of the world will be focused on this event to see which nation has the best unarmed combatants.

Although the U.S. Navy stole the prized team title, Bobby Clayton of the U.S. Army was clearly the audience's favorite contestant. Clayton won the gold medal in the lightweight division with a perfect "six wins, zero losses". Both Baldwin and Kubotsu feel that Clayton is an outstanding athlete who should be given an opportunity to compete in the CISM Taekwondo Championships. As one light weight contestant put it, "Clayton is one tough soldier, a human fighting machine."

### Results: 1st PLACE WINNERS

Fin — none
Fly — none
Bantam — M. Sheffield (Air Force)
Feather — C. Hafer (Air Force)
Light — B. Clayton (Army)
Welter — M. Hampton (Navy)
Middle — three way tie:
    T. Evans (Navy)
    A. Cudnohlofsky (Navy)
    B. Harris (Air Force)
Heavy — J. Boatner (Air Force)

The photo shows B. Clayton of the Army on the right facing off with D. Shelly of the Air Force on the left. This was Clayton's only tough match as he nosed out Shelly three points to two points.

2nd US Military Taekwondo Championships

# CHAPTER 5
# THE NAVY:
## IT'S NOT JUST A JOB; IT'S AN ADVENTURE! 1987

1987 Armed Forces Team Korea

Thank God for the Information Age. However, Google wasn't always my friend, but what did become my best source of information was speaking with those who were there at what I refer to as Ground Zero.

During one of my many phone conversations with GM Luis De La Rosa, formerly of the USMC, I continued to learn more about the origins of the Armed Forces Taekwondo Team and its early pioneers.

It was GM De La Rosa who provided me with the phone number of Chief Petty Officer William D. Baldwin, the founder of both the Navy Taekwondo Association *and* the Navy Taekwondo Team.

Via Zoom Call, I was able to interview Chief Baldwin. This interview taught me how and, more importantly, why he created the Navy and Armed Forces Taekwondo teams.

**Zoom Call Interview with William D. Baldwin, Pioneer of the Armed Forces Taekwondo Team**

**Louis Davis**: Chief Baldwin, of all the information that I've gathered prior to my locating you and reaching out to you, it has led me to believe that you were the pioneer of what would later become the Armed Forces Taekwondo Team, as well as the creator of the Armed Forces Invitational, which later also became the Armed Forces Championship. What was your motivation to create a team representing the US Armed Forces to include creating the Armed Forces Championship? How did that happen?

**William D. Baldwin**: Well, I've been involved in Taekwondo for many years while on active duty. I thought that there should be a vehicle for active duty people to participate in Taekwondo. Back in the early years, I was competing on my own and it was always a struggle to try to get any kind of support be it orders, or funding to go to major Taekwondo tournaments. So, I just thought that there should be a military Taekwondo organization.

**Louis Davis**: What year did you set out to make that happen? What year did you begin to set things in motion?

**William D. Baldwin**: 1972.

**Louis Davis**: Shortly before the founding of the World Taekwondo Federation?

**William D. Baldwin**: Yes.

**Louis Davis**: So, it's safe to assume that you were offered no support during that time so, what did you do to gain support? How did you manage to make things happen?

**William D. Baldwin**: I was constantly going to military leadership, talking to them. Every time I came back from a tournament, I showed them the results of the tournament. Mostly it was just a lot of talking.

**Louis Davis**: Yeah, I can imagine. Just out of curiosity, can you elaborate on that a little more?

**William D. Baldwin**: For example, in 1973, I believe it was. I was in the station in Guam and wanted to go to Korea to compete in a tournament over there. I worked my way up to Commander of Marianas, the Admiral, and I tried to convince him to get me some orders to go, and they wouldn't do it. So next, I spoke to the MWR (Morale Welfare and Recreation) director there, and it just fell on deaf ears, so I went to Korea on leave on my own time and expense to compete.

In doing so, I won the Heavyweight championship, so when I returned to my duty station, I took the certificates as proof of my success. I revisited that cycle of talking to people, and they finally said, "Okay, if this happens again next year, we'll fund you." I wanted to go, and of course, I transferred to a new duty station the following year.

**Louis Davis**: And you had to start the whole cycle again, the entire process, am I right?

**William D. Baldwin**: Right. That's the thing about the military: when you're constantly getting transferred from place to place and constantly traveling as much as we do, any arrangements you may have had with the previous duty station do not follow you to the next duty station.

It doesn't have an accumulative effect with what you do. At that time, I was with the security department working as a command investigator. So, I worked hard to establish my reputation there with the command. In doing so, you get a little bit more credence with the people at the command level.

**Louis Davis**: More of a voice?

**William D. Baldwin**: Yes.

**Louis Davis**: That makes sense. And then, once you got that support, was it always in the form of permissive TDY (Temporary Duty Assignment)? When did you start receiving financial support from the Navy?

**William D. Baldwin**: Financially, I didn't start getting any backing until about 1985, and that backing came from the head of Navy Sports in Washington, D.C.

I went to D.C. and had several meetings with him. Finally, I sold him the idea of a Taekwondo team competing to represent the military. The big thing there was I had to show him Navy-wide participation, and that was when I first began getting the Navy team together.

**Louis Davis**: Where was the first Naval Taekwondo competition held?

**William D. Baldwin**: The first tournament was held at the Naval Amphibious Base, Little Creek, Virginia, and those athletes came to compete much the as I did in the beginning, on their own time and with their own money.

1987 Armed Forces Taekwondo Team

**Louis Davis**: You mean Little Creek, Virginia; that base there?

**William D. Baldwin**: Yes, that's where Navy Taekwondo was headquartered, out of Little Creek, Virginia.

**Louis Davis**: That's something I never knew. Speaking of which, I assumed that the Marines stood up around that same time. There is a gentleman I've heard about named Larry Spears. Did he participate in any of these invitational tournaments that you set up?

**William D. Baldwin**: Yes, he came in with the Marine Corps team. And I think, don't quote me now on the year, but I believe it was 1986. It was Sergeant De La Rosa, Sergeant Spears; it was about a six-man team representing the Marines that came in and competed there at Little Creek.

**Louis Davis**: Yes, I've seen a picture trophy that featured Spears and other Marines posing with him. He said the Marine team had won it at this competition. There is this big trophy; I remember seeing a picture of it on Facebook that commemorates that tournament. Can you elaborate a little more on that tournament? How many competitors were at the tournaments? How many divisions did you have, that sort of thing?

**William D. Baldwin**: If I recall there were about 70 to 75 competitors, each from different branches of the service. At that time there would have been eight black belt divisions, so eight different weight classes. It was primarily just black belt competitors.

**Louis Davis**: Wow, that sounds pretty awesome. I wish I could have been there to see what that kind of competition was like back then!

**William D. Baldwin**: The military tournaments were always really tough-fought tournaments because we had a tendency to fight round-robin. So, you had to fight everybody in your weight division.

**Louis Davis**: Hmm. So it wasn't a single elimination. You had to bring some heat.

**William D. Baldwin**: Yeah. I always understood the need for single elimination in big tournaments, but I never did like them because it does really demonstrate who the better fighter was. But in a round-robin tournament, you win by beating *everyone* in your division. Therefore, you gotta be tough, you gotta bring it if you really want to show everyone what you got.

**Louis Davis**: I quite understand, Chief, more than you know. So then, how were you able to get the US Taekwondo Union involved? Because it seems that the next part of the military team's evolution would be the US Taekwondo Union. So how are you able to get an audience with the governing body of the sport at that time? How were you able to convince them to support the program that you had created?

**William D. Baldwin**: I had meetings at that time with Dr. Dong Ja Yang who was the president of the governing body. Originally, when we started, the AAU (Amateur Athletic Union) was the national governing body for Taekwondo and then shortly after that they changed over to the US Taekwondo Union (USTU). Dr. Yang had given me a letter of appointment to organize the military Taekwondo program within the USTU structure. There was already an established Class A division for each branch of the service.

**Louis Davis**: Really?

**William D. Baldwin**: Yeah, that was mandated by the Congressional Sports Act.

**Louis Davis**: Whoa! The Congressional Sports Act really carries a lot of power!

**William D. Baldwin**: Oh, yes.

**Louis Davis**: Hmm, I'm gonna have to do a little bit of research on the Congressional Sports Act. So, is that what gave the Armed Forces their way into the USTU?

**William D. Baldwin**: Yes.

**Louis Davis**: That being said, how did you learn about the World Military Championships and when did you decide to field an Armed Forces team? What year did you initially plan to field a team for that event?

**William D. Baldwin**: I first learned about the International Military Sports Council (aka CISM) during my time stationed over in South Korea back in 1980. That's when the first CISM games took place.

**Louis Davis**: Did you attend that event?

**William D. Baldwin**: Yes.

**Louis Davis**: How many countries were present at the first one?

**William D. Baldwin**: Well, the first one was a small competition, and the US military team was comprised of service members from the US Army's Eighth Army Division. They completely supplied the team. I tried to get on that team, but the ream wasn't really open to anybody except Eighth Army personnel.

**Louis Davis**: Do you remember a gentleman by the name of Hawk Hawkins? Does that name ring a bell?

**William D. Baldwin**: No.

**Louis Davis**: I believe Mr. Hawkins may have competed in that event. That's what I've been trying to speak with him about to get some clarity. Chances are that might have been the tournament he'd gone to, but I could be wrong. He was competing in the seventies around the same time you were.

**William D. Baldwin**: I had some limitations back then with where I was working, and I wasn't always able to go out openly into the public and do things.

**Louis Davis**: Hmm. So, your security clearance level kind of limited you on what you could do and what kind of interaction you had?

**William D. Baldwin**: Right.

**Louis Davis**: Hmm. Security clearances are notorious for doing that, aren't they?

**William D. Baldwin**: Hmm. They sure are.

**Louis Davis**: When the US Taekwondo Union had allowed the Armed Forces to start fielding a team, how many people were you able to take with you to the National Championships at that time?

**William D. Baldwin**: The first time I took about five people.

**Louis Davis**: Were they all sailors, or were they from different branches of the military?

**William D. Baldwin**: Well, the first team was comprised of sailors, because I took a Navy team, and the Armed Forces as a whole wasn't really functioning at that time. The team developed as I got the US Military Cup tournament going in Las Vegas in conjunction with the Western Regionals.

**Louis Davis**: Military Cup, eh? So that was, in a way, a predecessor to the Armed Forces Championships?

**William D. Baldwin**: Yes.

**Louis Davis**: Do you think that was the tournament that Larry Spears won? Or was it a previous tournament?

**William D. Baldwin**: That was probably the one he participated in and won the competition. Like I said, that competition was held in Las Vegas; he and a few other Marines came to that tournament quite a bit (US Military Cup).

**Louis Davis**: Now that makes sense; they did have bases close by in California.

**William D. Baldwin**: Right. That was part of the reason for doing that tournament in Las Vegas. This would give the West Coast people more of an opportunity to come because it was really less expensive than the one all the way out to Virginia Beach.

**Louis Davis**: When did the Army start making its presence known? Let me phrase it this way: when did you first become aware of the three people that were representing the Army out of Fort Bragg?

**William D. Baldwin**: The original Army? That was with Master Sergeant Owens. There's a lot of people in the early days that came and went within the system that I was trying to, you know, develop, the organization. I tried to get anyone in the military involved in Taekwondo into the organization. Master Sergeant Owens came down to watch one of the military tournaments at Little Creek, Virginia.

**Louis Davis**: There were two gentlemen from Fort Bragg that I'm aware of that attended one of the battles at Little Creek. That was Raphael Medina and Pedro LaBoy out of Fort Bragg. When did you first come into contact with those two?

**William D. Baldwin**: I believe that was in '87 at the Armed Forces team trials.

Louis Davis: Was that for the team that went to Korea later that year?

**William D. Baldwin**: Yes.

Louis Davis: Is that also when you introduced them to Bruce Harris? How did that whole conversation take place? Was it due to their success? Or should I say, Mr. Medina's success at the tournament? And his success reflecting greatly upon your handiwork?

**William D. Baldwin**: Yeah. You know, the team trials were actually a training camp. They came under orders to train with me at Little Creek. I believe that it was two or three weeks where we trained, and then at the end of that training camp we had a fight-off, and after the fight-off I selected the team that was going to go to Korea to fight.

Pedro La Boy was a Heavyweight. As for Medina, I don't remember his weight division, but his was a smaller weight division. Yeah, they were good fighters.

Louis Davis: Mr. Medina was a Bantamweight from the research I've done thus far.

**William D. Baldwin**: I wouldn't argue with that.

Louis Davis: No, sir.

**William D. Baldwin**: Actually, before that, I had taken a Navy team down to Fort Bragg, and we set up a match between the Army and Hampton Rhodes Military with Pedro Laboy.

Louis Davis: Wait a minute. I have pictures of an event similar to this, and I think the helmets you guys used had a headband with a, with a pad directly in the back of the head. Is that the same competition?

**William D. Baldwin**: I don't remember.

Louis Davis: Very interesting. I had no idea.

**William D. Baldwin**: Yeah. The team that I took down consisted of Navy sailors, one Marine and one Air Force airman. Because of all the events that I did, and at Little Creek, I opened it up to all the military. It wasn't exclusively Navy and that's where we started trying to blend all of the services into one organization.

We had a meeting in Las Vegas in 1984 where we actually established what, at that time, was called the Armed Forces Taekwondo Committee, and the name later changed to US Military Taekwondo Federation.

Louis Davis: Okay, who was present at this meeting?

**William D. Baldwin**: Myself, Daryl Kubotsu and Sterling Chase from the Air Force. There was an Army representative present but I don't remember the representative's name; all four branches were represented.

It was initially presided over by a representative from the United States Taekwondo Union and at that particular meeting is where I was elected as chairman. That meeting was the actual birth of US Military Taekwondo.

**Louis Davis**: Wow! Was this in 1986?

**William D. Baldwin**: This was in 1984. That's when the US Military Cup came into existence. The tournament we had was the US Military Cup and the team who won the trophy from that competition had to bring it back the following year and defend it. And that was the trophy I mentioned earlier; I still have it here in my house. It is an actual cup on a base. It's got a plaque on the front of it for the branch of service winning the cup each year.

**Louis Davis**: You wouldn't happen to have that close by with you. I'd love to see this.

**William D. Baldwin**: Sure do. Hold on.

**Louis Davis**: Wow!

Chief Baldwin goes off-camera and brings back a trophy that has several smaller plaques on it.

**William D. Baldwin**: This is the base. I don't know what you can see.

**Louis Davis**: Yeah, I can. I can make out parts of it. Wow!

**William D. Baldwin**: And then, of course, the cup went up on top of it. The cup is packed away but I still have it. Yeah.

**Louis Davis**: Damn. Which teams won the cup and which years did they hold the cup?

**William D. Baldwin**: The first one was 1985. 1986 was won by the Navy. 1987, the Air Force and 1988, Air Force. There was a guy from down in Panama named Gonzales. In 1989, there was a Marine named Sergeant Roberts who was the team captain. That was the guy that Sergeant Spears was with. I believe his name was De La Rosa. That's probably the big trophy from that year that you're seeing. In 1990 it was the Air Force, won by Bradford Carter.

Armed Forces Invitational Trophy

**Louis Davis**: Brad Carter?

**William D. Baldwin**: Yeah.

**Louis Davis**: What year was this?!

*One Team. One Fight. One Family.*

**William D. Baldwin**: In 1990.

**Louis Davis**: Oh, my God! Whoa!

**William D. Baldwin**: 1991 was the Air Force and Carter won again while he was stationed at Mountain Home Air Force Base; '92, Navy, '90, Navy, '93, Navy.

**Louis Davis**: Who won it those years for the Navy?

**William D. Baldwin**: Normally the name of the winning team is placed on the cup, not the individual. It was usually the team captain's name. '94 – '96 was when the cup was retired because the Navy had won it so many times. They had a very strong, powerful team: Mike Delgado, Eric Hampton, and Luis Rhoda.

**Louis Davis**: Luis Rhoda? I found him, by the way.

**William D. Baldwin**: Luis Rhoda? Down in Florida? Cuban?

**Louis Davis**: Yep. That's him, yeah. What about Ron Berry? Was he a part of that group?

**William D. Baldwin**: He came after that.

**Louis Davis**: What year did Ron Berry begin to work with you?

**William D. Baldwin**: Yeah, Ron began working with me around '90, I guess. I don't remember exactly what year he came onboard; the years may flux a little bit there.

**Louis Davis**: When did he leave the team?

**William D. Baldwin**: Around 2000.

**Louis Davis**: Wow!

**William D. Baldwin**: Ron used to always come up to the tournaments at Little Creek. I later appointed him as President of the Navy Taekwondo Association after I retired. I thought it should be an active-duty member that should be the president, so I chose Ron. There was a Lieutenant named Steve Goad that was president for a while before Ron was president; he was succeeded by Lieutenant Gail Johnson, who later, was Captain Gail Johnson.

**Louis Davis**: Boy, did I tumble down the rabbit hole with this! I am glad I've had the opportunity to reach out to you. I wanted to talk a little more about CISM and about the team that you sent to Korea in 1987.

**William D. Baldwin**: You'd probably like to see the big team photo, too.

**Louis Davis**: Yeah, if you have it.

**William D. Baldwin**: Yeah, I do.

**Louis Davis**: Please. I'd be honored!

Chief Baldwin goes off-camera, returning with a small picture with the 1987-Armed Forces Team.

**Louis Davis**: My first question is: now that you had some backing, how are you able to field this team? Was it all Navy sports, or was it Armed Forces Sports that funded this team to go to Korea?

**William D. Baldwin**: To attend the CISM Games? Yeah, it was Armed Forces Sports.

**Louis Davis**: Okay, how did that come about, who did you meet with? What happened?

**William D. Baldwin**: I met with the Navy Sports office, and they informed me that the Army was going to select the team out of Korea. So, I started doing some lobbying to keep that from happening and convince the Navy Sports office and the Armed Forces Sports Coordinator.

So, you're talking about several trips to D.C., constantly meeting with the people that were in charge, and finally I convinced them that there were enough active-duty Taekwondo participants here in the States to successfully host a training camp and the team trial. A couple of the guys that came to camp did come out of Korea from the Army, but I can't remember their names right now, though.

**Louis Davis**: That's okay. Names might not be that important. But from what I learned after reading some of the things that Rafael had written on his now defunct website, there were people from various fighting styles, not just Taekwondo, at the Armed Forces trial camp, vying for a slot on the team. Is that true?

**William D. Baldwin**: Yes, but none of them made it.

**Louis Davis**: Yeah?

**William D. Baldwin**: Well, their styles just weren't compatible to the rules. One of the first things I had to do at the training camp was actually teach them the rules via a referee seminar. This seminar helped them become aware of the rules that they had to compete by and during the competition.

We also had some independent styles that were all supposedly Taekwondo, but they were from independent groups. Many of those groups were doing Japanese-style point fighting and, of course, that was not compatible with Olympic-style rules.

**Louis Davis**: Yeah. Olympic-style rules are full contact and usually, if you screwed up, you hit the pavement.

**William D. Baldwin**: Well, some of them were wanting to punch to the head, which didn't work well. So, at the initial part of the camp, I separated into two groups: those that knew the Olympic style fighting system and rules, and those that didn't, and then I brought those that didn't up to speed. The general workouts would all be together, but some specific training was needed for those that weren't really aware of the WTF system.

**Louis Davis**: How long were the practices, and how many practices did you have per day?

**William D. Baldwin**: We did three per day. We did a morning session before breakfast consisting of the team going for a run in the morning. Then they went back, had breakfast, had about a two-hour break, came back into the gym, then a session before lunch, and they came back at two o'clock and did the afternoon session. And there was also a quasi-fourth session, which was more individuals working on things they needed to go over, focusing on particular strengths and weaknesses. So, it was a full day and they put in eight hours a day.

**Louis Davis**: And I'm almost certain the building smelled like Bengay for about a week.

**William D. Baldwin**: The Rockwell Hall Gym was where we trained. That was a very large gymnasium. It's big enough. It had six basketball courts in it.

**Louis Davis**: NICE!

**William D. Baldwin**: The big gym used to be a seaplane hangar and we had the whole west end of the gym, and I had bags hung around that area, and a circuit. It was this feature that all the guys noticed real quick when they first came to the camp. You see, I had all these big metal garbage cans sitting around the floor, and very soon they found out what they were for.

**Louis Davis**: Okay, what were they for?

**William D. Baldwin**: To go throw up in them. During the training sessions you throw up and get back in, and you keep going. It was a tough camp.

**Louis Davis**: My God! You were tough as nails.

**William D. Baldwin**: Yeah, there were the guys that lost thirty pounds. During the introduction, the message went out when they were applying for the camp, the message stated that "this is not a camp for beginners; this camp is for advanced competitors only! You are required to come to camp in shape and ready to go." Some of them did but quite a few of them didn't.

**Louis Davis**: Mr. Boltz stated the same to us during my years at Fort Indiantown Gap; that message continued my first year with the Army team in 1997. The standard was the same; you had to come to camp in shape. Now I understand the source where that standard came from. Again, that was your handiwork.

**William D. Baldwin**: Well, when you're going to do a fight-off and you gotta fight everybody in your division.

**Louis Davis**: You really whipped these boys into shape.

**William D. Baldwin**: Each fight: three rounds, three-minute round-robin.

**Louis Davis**: Yeah, I started off that way: three rounds, three minutes long, and if you burn your reserves in the first round, it's going to be a long match!

**William D. Baldwin**: And due to round-robin style fighting, the fight-offs took three days.

**Louis Davis**: That's what Master Medina said as well, that the round-robin fight-offs were three days long.

**William D. Baldwin**: Yeah.

**Louis Davis**: Hmm. How did you mitigate the injuries during the training and during the round-robin?

**William D. Baldwin**: If there was an injury that stopped you from fighting. You were out.

**Louis Davis**: Go hard or go home, eh?

**William D. Baldwin**: That's right. I come from the old school!

**Louis Davis**: Oh, I saw what those *hogus* looked like back in the day. They had strips of bamboo on them, and then the ones that succeeded them. I was told by friends of mine who remember using them, if your hands and feet weren't conditioned, it's going to be a brick passer, am I correct?

**William D. Baldwin**: Yes, that's correct. I think it was 1973. I was fighting in a Korean championship called the Cherry Blossom festival in Jinhae, Korea. We used those *hogus* that you mentioned, and the bamboo strips had been beaten to sawdust!

**Louis Davis**: Yikes! Splinters! I'm trying to imagine this, and I've got no reference. I've seen those *hogus* only once during my teen years here in Minnesota. There was a gentleman by the name of Byung Yul Lee and when he was teaching the system, he had two of those chest protectors in his dojang. One of them had a white background and the center was colored red, and then the other one's center was colored blue. Both had those small strips of that bamboo laid over them. People who remembered using them told me stories about sparring with those *hogus* and how the bamboo conditioned your hands and your feet.

**William D. Baldwin**: Yeah. Well, when I first started fighting in Korea, we didn't even use *hogus*; it was optional.

**Louis Davis**: Knuckle to bone?

**William D. Baldwin**: Yeah.

**Louis Davis**: That explains things. You guys were pretty badass back then!

**William D. Baldwin**: Part of the original rule, why we didn't punch to the face. It just did too much cosmetic damage.

**Louis Davis**: So, from the looks of it, your fighting styles were based loosely upon the Kyokushin Karate style of sparring. That being said, during the fight-offs at Little Creek, was there anyone there taking video footage to help you decide who was going to make the team and who didn't?

**William D. Baldwin**: Yes, I had video cameras set up and I went to each ring. I had a whole chest full of video footage, and that's part of what we did in the evenings; we would go over the videos.

I also let the team guys take the videos over to the barracks where they would sit and watch them, and this helped them to better understand the sport. There were a lot of comments about the fighting over there. A lot of the guys that came to the camp that had never really experienced WTF (World Taekwondo Federation) style fighting were very impressed by the fighting. I still have all of those videos today. They're old VHS tapes.

**Louis Davis**: Oh WOW! I can't wait to get those converted into MP4 format.

**William D. Baldwin**: Yeah, that is if they haven't deteriorated. I haven't looked at them in years, but there's probably 30 or 40 videos at least.

**Louis Davis**: I have got to get my hands on this stuff. This is valuable.

Let's talk about Korea and the CISM games and what happened once you guys got over there. I've seen some pictures that Master Medina had furnished of your time there. So, what was the mindset of the team once it was selected?

**William D. Baldwin**: The team was all gung-ho, ready to go!

**Louis Davis**: I understand. You guys took three medals that year.

**William D. Baldwin**: I don't remember how many, but we finished fourth overall.

**Louis Davis**: It was Melvin Boatner, Tim Hightower, and Rafael Medina. They all took bronze medals from what I saw on the website taekwondata.org.

**William D. Baldwin**: They were really prepared for the arena, and this was probably the largest and most significant tournament they had ever been to. It was in a large arena and there were thousands of people watching, and there was even television coverage.

**Louis Davis**: Television coverage?

**William D. Baldwin**: Oh, yeah, The Mosan Broadcasting Company (Korean Broadcasting) was broadcasting it live on local Korean television. The competition was held in a newly built facility that was built to support the '88 Olympics called Jamsil Arena. So, this was like a pre-Olympic event.

**Louis Davis**: Wow! Was there anyone from the Armed Forces Network, or from *Stars and Stripes* or any of the US military media outlets; anyone from those military news outlets covering you guys during the competition?

**William D. Baldwin**: Unfortunately, no. I was really surprised when we first got there. I had an interview with a representative from the AFKN which is Armed Forces Korean Network, and they didn't seem to have much interest.

**Louis Davis**: Wow! Unbelievable! So, who was the team captain for that group? Who did you select as team captain?

**William D. Baldwin**: Sandusky, Lieutenant Sandusky.

**Louis Davis**: Which branch of service was Lieutenant Sandusky from?

**William D. Baldwin**: The Navy. He was a back seater for a fighter (R.I.O.).

**Louis Davis**: If he was a back seater, it's safe to assume that he was in an F-14 tomcat?

**William D. Baldwin**: Yeah.

**Louis Davis**: Who else assisted you in coaching the team and getting the team ready? Who was your assistant?

**William D. Baldwin**: My assistant coach was Bruce Harris from the Army.

**Louis Davis**: Would you care to comment about him? How did he end up getting the position?

**William D. Baldwin**: I appointed him to the position. He was very gung-ho. Although his official title was "trainer," he helped me in all the aspects of running the camp.

**Louis Davis**: So, he was able to cover down on some of the gaps that you encountered during that time?

**William D. Baldwin**: Right.

**Louis Davis**: Who did most of the coaching during the event; was it you or was it Coach Harris?

**William D. Baldwin**: I did the bulk of it while Coach Harris covered down on some of the other matches.

**Louis Davis**: When did you officially step down as a head coach for the Armed Forces team and why did you choose to do so? What happened?

The 1986 World Military Championships (CISM) Provided by Rafael Medina

**William D. Baldwin**: Well, shortly after the team came back from CISM in 1987, the position went away and there wasn't an ongoing position as coach for the Armed Forces team.

So, each time they did an event, they'd appoint a coach.

For example, after we came back from Korea, I would stay for about a month at Little Creek writing and reviewing the after-action report, and the follow-up, and then it was over with.

**Louis Davis**: This report was suggesting to the Navy that Armed Forces Sports would have wanted to keep this going?

**William D. Baldwin**: They were an interesting group of people. They'd just gotten burned by the bobsled team.

**Louis Davis**: A bobsled team?

**William D. Baldwin**: Yeah, the Navy had a full-time US Olympic bobsled team, which was also headquartered at Little Creek. They competed at Sarajevo, and they did not qualify.

**Louis Davis**: They didn't qualify? Oh no!

**William D. Baldwin**: They did not, and they were immediately disbanded.

**Louis Davis**: Wow! They didn't even accomplish the mission.

**William D. Baldwin**: If you're familiar with other sports, there were other sports where they (the Navy) had invested a lot of money, and then it didn't pan out. So, they were not inclined to make long-term investments in different sports.

**Louis Davis**: Did the Navy have an equivalent to the Army's World Class Athlete program at that time?

**William D. Baldwin**: Yes, they did.

**Louis Davis**: When was that established for the Navy?

**William D. Baldwin**: That goes all the way back to what's called All Navy Sports. I don't know how far back it goes but it's been around a long time back.

**Louis Davis**: Was there a program like this for the Navy for Taekwondo?

**William D. Baldwin**: Not really but the Navy relied on me to do it and I was running a program, and it was a quasi-program, as-events-came up-type-thing.

**Louis Davis**: I see. Who succeeded you once you stepped down? Who took over? Who kept things going?

**William D. Baldwin**: Basically, no one. That's really why the program (Navy WCAP/ Elite Sports Program) went away.

**Louis Davis**: What the hell? I didn't see that one coming.

**William D. Baldwin**: One of the questions you've asked me regarding one of the elite athletes the Navy Taekwondo Team had, I'm trying to think of her name. She and her husband fought for me on the Navy team.

**Louis Davis**: Elizabeth Evans?

**William D. Baldwin**: Yes. Elizabeth was in the Elite Sports Program.

**Louis Davis**: When did she first join the team? Did you do any work with her?

**William D. Baldwin**: No, I didn't.

**Louis Davis**: I know of her husband, Troy. So, did Troy Evans fight for you as a member of your team?

**William D. Baldwin**: Yes, that's correct.

**Louis Davis**: The details of how Liz got involved in the sport are very sketchy, at best, and I don't have much to go on; most of it is based off what people could tell me about her reputation as a competitor, and I can tell you it was well deserved. My stomach still hurts now from my holding a chest protector while helping her warmup during the World Military Games in 1999. But I know nothing else about her.

**William D. Baldwin**: I don't really either because she came in, basically, as I was going out. I had some heart issues and after about '88 I really tapered off because one, I was retiring and two, the medical issues. It had really slowed me down there for two or three years.

That's when I started appointing the different Navy people as president of the Navy Taekwondo Association. The Air Force, Army and the Marines had their associations up and going and I had been serving as an umbrella

Rafael Medina receiving his medal,
2nd CISM, Korea

Major George Nobles was heading up the Marine Corps. He was there at Quantico, at Headquarters Marine Corps. Daryl Kubotsu was Air Force, and then he kind of passed it off to Sterling Chase and then the Army, after '87 really started doing their own thing.

That's when they started getting Indiantown Gap up and running and Bruce Harris took over as head coach for the Army. The Army made a play with the Armed Forces Sports to take over as the head coach of Taekwondo.

**Louis Davis**: Hmm! Do you think that was a wise idea or do you think that it could have gone a different way?

**William D. Baldwin**: I think it could have gone a different way. But there was a lot of the old military rivalry between the Army and Navy, that sort of thing. At the forefront of it was Bruce Harris.

*One Team. One Fight. One Family.*

**Louis Davis**: I can understand that. Is there anything else that you wish to add to this interview?

**William D. Baldwin**: Not that I can think of offhand.

**Louis Davis**: Chief Baldwin, this has been an honor and a privilege to finally meet you and to get your side of things to actually record it. When we first spoke by phone, I wished that I would have had something connected to my phone to where I could have recorded our conversation the first time around. Again, this has been an honor.

**William D. Baldwin**: Any of the source material I have you're welcome to it. There's stuff boxed up and put in the attic that I haven't seen for 30 or 40 years, both photos and videos. You're more than welcome to it because it doesn't do me any good.

**Louis Davis**: Well, I would like to start with some of the pictures that you may have, and most definitely the video footage. It will take me a while, but I can get that footage converted to a modern-day format MP4 where you could watch that on your computer.

Pretty much, that's the end of it. Oh, there is one other question I want to ask you. You mentioned Daryl Kubotsu and Sterling Chase. I've tried to find Mr. Kubotsu via Google search, and I've come up empty. Have you maintained any contact with either Mr. Kubotsu or Mr. Chase?

**William D. Baldwin**: Daryl seemingly fell off the edge of the earth a few years ago, and I have no contact with him. The last dealings I had with him, he was out in California, and he was working at Now Care at two or three different locations out there. He's a physician's assistant. So, the last few years I don't know what has happened with him. Our mutual instructor, Grandmaster Dong Sup Kim, has passed away so like I've said, I don't know what's happened with him. Sterling is out in Colorado Springs.

**Louis Davis**: What? This guy was right under my nose!

**William D. Baldwin**: Yeah, while he was still with the Air Force he worked there at Cheyenne Mountain.

**Louis Davis**: NORAD? Oh, I know that place very well. Alright, Chief Baldwin. I have no further questions, Sir. Thank you for all of this, Chief. Have a great evening, Sir, and Merry Christmas!

**William D. Baldwin**: Same to you.

1987-Armed Forces Taekwondo Team
The accumulation of CPO Baldwin's efforts to get Taekwondo recognized as a sport.

Memories of Rockwell Hall Gymnasium

Memories of Rockwell Hall Gymnasium

Memories of Rockwell Hall Gymnasium

Memories of Rockwell Hall Gymnasium

Memories of Rockwell Hall Gymnasium

THE NAVY

# HOORAH, HERE COME THE MARINES!

The USMC Taekwondo Team

Initially I intended to share the interview with one of the pioneers of the Marine Corps Taekwondo Team. Unfortunately, the recording containing the interview with Luis De La Rosa was damaged and unusable.

GM De La Rosa, if you're reading this, I humbly apologize to you, Warrior, for not being able to share your story. You, GM Spears, and your accomplishments as fighters of the USMC have *not* been forgotten.

At around the same time that the Navy Taekwondo Team was stood up in the mid-1980s under Marine Corps Major George Nobles, two Marines, Larry Spears and Luis De La Rosa, stepped up to represent the Corps. True to their heritage as Marines, they've left their mark as fighters during the days of the US Armed Forces Invitational, and GM Spears became the first Marine to win the cup for the Marines.

# CHAPTER 6
# THE AIR FORCE:
## "AIM HIGH, KICK LOW!"

1995 USAF Taekwondo Team
I have searched high and low for old team photographs of the Air Force Taekwondo team from the '90s era. I was able to find a photograph of the team from 1995. The team was headed by Coach James Arrington. Brian Johnson, Keith Young, John Holley, Curtis Brown, and David Arrington.
Photo courtesy of Brian Johnson

I have searched high and low for old team photographs of the Air Force Taekwondo team from the '90s era. I was able to find a photograph of the team from 1995. The team was headed by Coach James Arrington. Brian Johnson, Keith Young, John Holley, Curtis Brown, and David Arrington.

Following Chief Baldwin's interview, I immediately took to the internet in an attempt to reach Daryl T. Kubotsu, the pioneer of the Air Force Taekwondo team. To my disappointment I was unable to locate an up-to-date address and/or phone number.

However, all was not lost. Following the suggestions of Chief Baldwin, I returned to Google to locate GM Sterling Chase Sr. By God's grace I was able to locate him via the following website: www.chsmartialarts.org/home/instructors-bio. The website also included a phone number which I promptly, eagerly called and left a voicemail.

Within the next few hours, I received a call from GM Chase who was willing to be interviewed via Zoom.

**Zoom Call interview with GM Sterling Chase Sr., Pioneer of the Air Force Taekwondo Team**

**Louis Davis**: Grandmaster Chase, it is an honor and a privilege to be able to interview you for the Zoom call. As you know, I've been quietly gathering data, facts, and information on the history of the Armed Forces Taekwondo Team. I assume that you're well aware that I myself was once a member of the Armed Forces team representing the US Army. So, if you don't mind, we'll just get right into it.

**GM Sterling Chase Sr.**: Before we begin, I wanted to let you know that I just got off of the phone with a couple of the other people that were instrumental in the Air Force Taekwondo Team back when I was active in the sport.

Each of them has stated that they're willing to be interviewed. One of them is Grandmaster Jay Dunston who took over as the President of the Air Force Taekwondo Association when I retired. He was formerly the Alaskan State President for USTU as well.

**Louis Davis**: Really? When was this?

**GM Sterling Chase Sr.**: In 1988.

**Louis Davis**: This is great information. I have a few questions that I'd like to ask you regarding how the Air Force team got its start. So, my first question is exactly when was the Air Force Team founded and who were its early pioneers? What was the motivation for forming this team?

**GM Sterling Chase Sr.**: Well, that happened as I came on board in '86. I received a call from Daryl T. Kubotsu; he was a major in the Air Force and he knew some of the people that I knew and since I was stationed here at Cheyenne Mountain (NORAD), he asked me to be his eyes and ears, because he couldn't always make the (USTU) board meetings. So, at that time back in 1986 was when I registered the Air Force Taekwondo association under the USTU.

**Louis Davis**: So, who else were the other Air Force pioneers?

**GM Sterling Chase Sr.**: Back then there were a lot of athletes that were competing that wanted to officially compete underneath the Air Force flag, but because the Air Force Sports office would not recognize Taekwondo as a sport, we couldn't do that, unfortunately. A lot of the athletes I didn't know very well, some of them I did.

But because I was more into the political side of the USTU, I worked to create the Air Force Association and we were the focal point for Air Force members that were wanting to compete and be on the US Team.

**Louis Davis**: I see...that sounds like a monumental task. So, aside from yourself and Major Kubotsu, were there any additional airmen who helped to get the team started?

**GM Sterling Chase Sr.**: No.

**Louis Davis**: So, it was just the two of you that pioneered the creation of the Air Force Taekwondo Team and its Association under US Taekwondo Union?

**GM Sterling Chase Sr.**: Well, back then we mostly had just the athletes. Nobody wanted to get into the politics or deal with these guys, because of the fact that everybody that was competing was either first Dan or second Dan and all the guys on the political side of the house were 6th Dan, 7th Dan, 8th Dan.

Because of our low belt rank they (the USTU politicians) were usually 6th Dan – 9th Dan Masters and most of them wouldn't listen to us, or anything we had to say or offer. So, it was pretty much on my shoulders, because I was here in Colorado Springs, and most of the board meetings were held here.

**Louis Davis**: So, is this why you chose to get into the political side of things?

**GM Sterling Chase Sr.**: Yes.

**Louis Davis**: It makes sense. In doing so, what challenges did you face getting the team established, to include the challenges to get funding from MWR and Air Force Sports? Can you delve into what were those challenges?

**GM Sterling Chase Sr.**: The goose egg was that the Air Force Sports office did not want to recognize Taekwondo as a sport, and due to that our airmen were not eligible to apply for permissive TDY as a means of support. And that became my biggest goal, to gain the necessary support for the team, especially being here in Colorado. I worked very hard to gain that support any and every way that I could. I took any avenue that would help get the Air Force team recognized by the Air Force Sports office. We just needed them to recognize our sport.

Once the sport was officially recognized by Air Force Sports, our airmen could then go to their commanders and request permissive TDY to compete representing the Air Force. When this happened, they no longer had to use their vacation time (leave) to compete. We weren't looking for funding just yet, we were just looking to be recognized by the Air Force Sports Office.

I mean, Taekwondo was getting big and prior to getting the necessary support from Air Force Sports, we were doing everything on our own time with our own individual funds. We had to do things this way because Air Force Sports didn't recognize Taekwondo as a sport at that time.

**Louis Davis**: Were you given time to train when you qualified for, say, the state championships and the Nationals?

**GM Sterling Chase Sr.**: No, all of that was done on our own time. So, prior to gaining the recognition and support from Air Force Sports, we were forced to use our vacation time to train and compete. At that time our commanders would not give us any additional time nor financial support.

**Louis Davis**: So, what year did the Air Force Team attend the first USTU National Championships? How many people represented the Air Force at the Nationals?

**GM Sterling Chase Sr.**: As a group? They had attended it prior to me coming on board. It wasn't an Air Force Taekwondo Association; it was just Air Force guys that qualified through whatever state that they were stationed in. The Air Force guys were like "Well, if the Army did the same thing, we should too."

Navy was the first branch of service to sponsor a Taekwondo team and that was due to a Grandmaster Baldwin (CPO Baldwin US Navy), but it was many years ago.

**Louis Davis**: Which year did the Air Force show up at the USTU National Championships as a group?

**GM Sterling Chase Sr.**: As a group? I know that they initially showed up in 1985 but that was before I came on board. In 1986 when I came on board they were like, yeah, we had four guys who competed at the Nationals, but we couldn't walk under the Air Force flag as an official Air Force Taekwondo team because Air Force Sports didn't recognize us.

USTU recognized the Air Force Taekwondo Association; we were allowed to attend meetings as members of the Air Force but not as official representatives of the US Air Force.

It was the Air Force Taekwondo Association that gave us a voice during board meetings, and I was able to sit in these meetings as a member of the Air Force but not as an official representative of the Air Force.

I attended all of the board meetings and stuff, and I would take all the information that I gathered from these meetings and forward it to Randolph Air Force Base. What they did with it afterwards, I have no idea.

**Louis Davis**: Speaking of challenges, what challenges did you face getting the team established with the current governing body at that time (USTU)? Was it an easier approach with USTU than with Air Force Sports, or did you have more hurdles to go through?

**GM Sterling Chase Sr.**: Well, they (USTU) put up the hurdles but it was a much easier approach because after going through the US Olympic bylaws, I learned that we in the armed forces are what is considered group B members of the USOC, and being as a group B member, a group A member cannot deny us entry into the program; to do so would cause them to lose their Olympic status. So, groups such as AAU, YMCA would be considered group B members by USOC.

So, when it came to elections, we had a vote. As the Air Force Taekwondo Association we never claimed to speak on behalf of the Air Force, however we were speaking on behalf of Air Force athletes, and that's one thing I made very clear with the USTU and Randolph Air Force Base on any position that I was in and any meeting that I attended, that I was not representing the Air Force nor Air Force Sports, I was speaking on behalf of Air Force Taekwondo athletes.

**Louis Davis**: You've mentioned Randolph Air Force Base. Was this the home of Air Force Sports?

**GM Sterling Chase Sr.**: Yes, Air Force Sports was located there.

**Louis Davis**: Oh, okay, at first, I thought they were in D.C. like Army Sports.

**GM Sterling Chase Sr.**: I don't know about now because I retired in 1997.

**Louis Davis**: Did the Air Force Taekwondo team have its own WCAP (World Class Athlete Program)?

**GM Sterling Chase Sr.**: WCAP? Yeah, but that program was established after I'd retired. So, any questions about Air Force WCAP moving forward, I don't know anything more about them.

**Louis Davis**: Do you remember what year the team was officially recognized by the Air Force? Rough guesstimate?

**GM Sterling Chase Sr.**: A rough guesstimate is 1988 because Grandmaster Jay Dunston was paid TDY to officially attend the 1988 Nationals. He is the only one that I know of that can confirm that he was paid TDY at that time.

**Louis Davis**: Okay, which leads me to my next question, and it's regarding the 1987 Armed Forces Championship in Little Creek, Virginia, the one hosted by Chief Baldwin. Did anyone from the Air Force attend that event that you were aware of? Do you know who those athletes were?

**GM Sterling Chase Sr.**: Dwayne Harris was one of them. In fact, I just got off the phone with him today, earlier today, and he confirmed that Melvin Boatner and Timothy Hightower also were there from the Air Force.

**Louis Davis**: Melvin Boatner and Timothy Hightower?

**GM Sterling Chase Sr.**: Correct. So, those three were there.

**Louis Davis**: So, the Air Force *did* have representation! Outstanding! So exactly when did you first come into contact with Chief

Baldwin? When did these guys first learn of the invitational?

**GM Sterling Chase Sr.**: I knew Baldwin through the Western Regionals tournament. It was because of Grandmaster Dong Sup Kim. GM Kim was from Las Vegas, and he was the Nevada State President for USTU.

He hosted a competition called the Western Regionals, which originally started as a three-day tournament, and he attached the Armed Forces invitational to the morning portion of that tournament.

This portion of the tournament gave all members of the US Armed Forces an opportunity to compete against each other, however this competition is not sponsored by any of the branches of service. This competition was sponsored by GM Dong Sup Kim, and he provided the medals for the military competition.

**Louis Davis**: When did this competition take place? What year was the first competition held?

**GM Sterling Chase Sr.**: The first one that I attended was in 1985. I had just returned from Iceland when initially I heard about it so, I decided to go down and see for myself. I also took a couple of students from Peterson Air Force Base with me.

It was at this competition that I first met with Baldwin and Kubotsu and all of them, and learned about the team he was working to create.

**Louis Davis**: And the rest, as they say, was history?

**GM Sterling Chase Sr.**: Oh, yes. Now Grandmaster Dwayne Harris, who was also part of the 1986 Armed Forces team, he was Grandmaster Kim's number two guy. He was chosen to run those Western Regional tournaments and he was still active during that time, because he was stationed at Nellis Air Force Base.

**Louis Davis**: So, he was ideally placed then?

**GM Sterling Chase Sr.**: Totally.

**Louis Davis**: The Air Force athletes who attended the Little Creek competition, their names were Hightower and who was the other gentleman?

**GM Sterling Chase Sr.**: Boatner, Melvin Boatner.

**Louis Davis**: Did either of those guys go on to medal at the World Military Championships?

**GM Sterling Chase Sr.**: They did. I remember what place they both got. In fact, I found it in the *CISM Magazine*; it's featured in the 74th edition and lists all the medalists from that year.

**Louis Davis**: Did that magazine include Rafael Medina from the Army?

**GM Sterling Chase Sr.**: Yes, Medina was on the US Armed Forces Team, and we remembered how fast he was. I think he took bronze at the event.

**Louis Davis**: Rafael Medina, my senior belt, my senior Grandmaster, he was in the Army.

**GM Sterling Chase Sr.**: What? Was he really?

**Louis Davis**: He certainly was.

**GM Sterling Chase Sr.**: I was on the phone with Dwayne Harris just before this call, and I'm like, I remember him, and everybody was like, "Yeah, what branch was he in?" And we thought that he was a Marine. And then, of course, Dwayne and I thought "Yeah, he was that fast little guy," so we tried to remember which branches of service placed at CISM.

We only took three medals in the '87 CISM games. This was also the first CISM games that the Air Force sent athletes to, and Grandmaster Dong Sup Kim from Nevada was our official referee that went with the team.

**Louis Davis**: Oh, that's who that Korean master was in the team picture.

**GM Sterling Chase Sr.**: Correct, that's Grandmaster Dong Sup Kim and he was the highest-ranking international referee in the United States because his referee number was 100. I don't know his history right now, but he was one of our biggest supporters of the Armed Forces Taekwondo program.

**Louis Davis**: Who else was supporting the armed forces at that time?

**GM Sterling Chase Sr.**: Another one of our biggest supporters was Grandmaster Jerome Ritenbach. He was retired Army. GM Ritenbach was one of the vice presidents of the USTU. He was one of the "round-eyes" to become vice president. *(Pause)* Did I say, "round-eye"?

**Louis Davis**: Yes, you did.

**GM Sterling Chase Sr.**: I went back into the fat of things, sorry.

**Louis Davis**: You said he passed away.

**GM Sterling Chase Sr.**: Yeah, he passed away quite a few years ago. He was living in California at the time. So, like I said, he was one of our biggest supporters because during one of the CISM games, while I was stationed in Korea that year, he called me up and I took leave and I was selected by him, a representative for the USTU along with him, seated at the CISM games for that season.

**Louis Davis**: What year was this?

**GM Sterling Chase Sr.**: It had to be between '90 and '92, because that was when I was stationed there for two years. That's a rough guesstimate.

Unfortunately, I don't have the exact dates. I got all that stuff in files.

**Louis Davis**: If you're able to scan some of that stuff and then email it to me, it would be greatly appreciated.

Mr. Chase holds up a document folder to the camera.

**GM Sterling Chase Sr.**: I don't know if you can see this. It's one of twenty that I have on

USTU. It's documentation of all the board members, and from all the board meetings, and every email that I received on behalf of the Air Force Taekwondo Association, and every page is laminated, all that kind of stuff. So, I've got it in my storage shed!

**Louis Davis**: Do you have footage of any of these fights?

**GM Sterling Chase Sr.**: No, I don't. I was the guy that walked around with the camcorder, and not take pictures. I was so into the moment I wasn't thinking "Hey I got this camcorder in my hand." Back then, camcorders were the size of a ghetto blaster.

**Louis Davis**: And I believe we've already covered the question regarding the World Class Athlete program. My next question is a little more centered around one of the Air Force's most prominent fighters, Brad Carter. Do you know exactly, when did he join the team?

**GM Sterling Chase Sr.**: I can't really tell you when he joined the team initially. Brad called me because he had heard about the Air Force athletes, and he was told to contact me. So, he and I had a lot of conversations, and I told him that I would do everything in my power to try and get what we need to get support to the athletes.

It was at that particular time when Bruce Harris had stepped into the game, setting up Indiantown Gap, Pennsylvania as a training site and as a focal point to be able to get onto the team. It was around that time that Brad Carter came. I can't really tell you who Brad Carter's instructor was, but we think that it may have been Bruce Twain from Vermont. Then there was Grandmaster Gossett.

**Louis Davis**: Who succeeded you as president of the Air Force Taekwondo Association?

**GM Sterling Chase Sr.**: After I had stepped down from the presidency of the Air Force Association, then Jay Dustin took over and I served as his vice president until he stepped down.

**Louis Davis**: Why did you step down? If you don't mind me asking, what was it? What's the deciding factor for retiring from the Air Force Taekwondo Association?

**GM Sterling Chase Sr.**: It was time to hang up the uniform. Well, I just retired from the Air Force in '97 but I was hanging in there trying to help everybody out. Secondly, because I did have the contacts, I was still getting a lot of the information from the board members or the Board of Governors, that's what we called it.

I knew a lot of the state presidents and a lot of the Grandmasters, so we always kept in communication; so, anything that I could do to help out the athletes. That's what I was there for because they needed a voice. We just didn't have one.

**Louis Davis**: How have you continued to contribute to Taekwondo and the overall Air Force program after retiring from the ranks?

**GM Sterling Chase Sr.**: Well, you know, when I found out that WCAP was training somewhere

here in Colorado, I made some calls to find out what their schedule was and thought that maybe I could just pop in and just say hello to some people and stuff, and everybody was pretty much closed door on that idea. So, I talked to a Navy guy that I knew from back in the day, you know I'm sitting there now. I can't think of his name to save my soul; I'm looking at him right now.

**Louis Davis**: Can you give me a description of him? I might know who you're talking about.

**GM Sterling Chase Sr.**: You know, God I'm just so bad with that. Everything's throwing me off today.

During New Year's I found out that I'm going to be a grandfather. That's got my mind completely messed up because I kept telling my kids that I'm not ready, and they're like, "Well, we're forty." I don't care. I'm still not ready. Oh yeah, so I'm having a hard time this week. But this guy, he was on the Navy team…and what is his name, the gentleman from Athletes Without Limits? This is the instructor.

**Louis Davis**: Luis Torres?

**GM Sterling Chase Sr.**: No, no, no, not Torres.

**Louis Davis**: Athletes Without Limits? The only other person I can think of might have been Johnny Birch Jr. But Johnny was not from the Air Force, he was Army.

**GM Sterling Chase Sr.**: Okay, yeah, the gentleman, and that's with the para-taekwondo for Athletes Without Limits…

**Louis Davis**: You said he was Navy.

**GM Sterling Chase Sr.**: No, his instructor was Navy.

**Louis Davis**: This instructor might have been Ron Berry.

**GM Sterling Chase Sr.**: Ron Berry! There we go. Ron Berry. So, I contacted Ron Berry a couple of years ago when I had found out about this because this was right after Covid.

**Louis Davis**: Hmm.

**GM Sterling Chase Sr.**: And I saw Ron Berry on something and then I reached out to him, and he called me a couple of weeks later. We had about a five-hour call, you know, just catching up on a whole bunch of stuff, because a lot of people that he wasn't in touch with knew me, and we were just cross-referencing my stuff. There were a lot of things missing and that was one of the issues. See, a lot of the athletes didn't really know about all the stuff that I was doing behind the scenes.

I was just trying to be a voice for them to ensure that they got the information and were consistently kept in the loop. So, they didn't see me on the floor at the tournaments, but I knew everybody. Plus being here at the Olympics Center, I trained with a lot of the '88 Olympic athletes for a couple of years.

The farthest I got was in 2000. I won the nationals, in both forms and fighting and then that was about all the competition I'd done personally because I was more of an administrator, and I was running schools. So yeah, my thoughts were "Let somebody else have fun."

**Louis Davis**: Well, what else would you like to add? So, I mean the floor is open. Anything else that you'd like to pass on?

**GM Sterling Chase Sr.**: Information. And so, with the couple of phone calls that I've made this afternoon, you know. We're all sitting there trying to remember names and stuff, and I know that once I get into my storage shed and go through those folders, there's a lot more I can pick out. So, Dwayne Harris and Jay Dunston both have your contact information, so if you have any questions that they can help out with. You know we've always been very supportive of each other.

And see, Grandmaster Harris fell under Grandmaster Kim for years, even after he retired, because he kept on running those Western Regionals until Covid. I haven't talked to Baldwin in years.

**Louis Davis**: I believe I can help you with that.

**GM Sterling Chase Sr.**: Yeah, yeah, that would be good too, because Perry was telling me that some of the guys were talking about getting a reunion together. There's regional.

**Louis Davis**: We've had a few. I'm sorry. Are you referring to Reginald Perry? You mentioned someone named Perry. Who is the one we were just talking about from the Navy? Berry, Ron Berry. There we go.

**GM Sterling Chase Sr.**: Who were we just talking about from the Navy?

**Louis Davis**: Berry, Ron Berry.

**GM Sterling Chase Sr.**: Yeah, because Ron had mentioned that. And I was like, okay? Well, you know, if you guys ever want to invite me you've got my contact information, because Bruce Harris knows where I'm at because when he retired, he moved here to the Springs. Bruce and I don't get along; he knows it and I know it and I'll just say that outright. I think everybody that knows both of us, they know we do not get along.

**Louis Davis**: We'll talk offline about that for the sake of the book.

**GM Sterling Chase Sr.**: Yeah, that sounds great. But yeah, so you know, it's not like I'm not easy to get in contact with. I'll just put it that way, 'cause I'm on Facebook. I'm on LinkedIn. I'm on Twitter. I'm on all of that stuff. So, it's not like I'm hiding; I've been there. So as far as continuing with Taekwondo, I'm giving back to Taekwondo. I've created a nonprofit which will be eight years in May.

I work with what we call adaptive athletes. So, I work with the special needs and the disabled. So, 90 percent of my student base are special needs and disabled students.

And that's what I've been working with on a regular basis for the majority of eight years. As for the nonprofit, I've been working with Move United, which is a national organization for adaptive athletes. At the AAU level we've had students qualify for the U.S. Team.

It could be global. So, with the new members, we're doing pretty good.

*One Team. One Fight. One Family.*

**Louis Davis**: Have you connected with Athletes Without Limits?

**GM Sterling Chase Sr.**: Oh, yeah, I know the director and the founder because he's on the board with Move United, and he knew us when we came in. We're the only martial arts listed in Move United.

**Louis Davis**: Do you know a Master Johnny Birch by chance? He's also a part of Athletes Without Limits.

**GM Sterling Chase Sr.**: Yeah, I know Birch. I keep thinking Barry Partridge.

**Louis Davis**: No, Johnny Birch Jr. Was out of Florida initially. He's in Pennsylvania now.

**GM Sterling Chase Sr.**: You're talking about the Navy guy?

**Louis Davis**: No, that's Ron Berry. Johnny Birch is a completely different person.

**GM Sterling Chase Sr.**: So, Ron Berry is his instructor.

**Louis Davis**: Hmm…I'll have to ask Master Birch about that the next time I speak with him.

**GM Sterling Chase Sr.**: November was the last time I communicated with him because we're getting our athletes registered with the Athletes Without Limits. My goal is to have fifteen of them registered with Athletes, so they can compete internationally.

**Louis Davis**: Once we're done with this call, I will call you back. I'll try to get Master—Mr. Birch on the phone and see if I can make that happen for you.

**GM Sterling Chase Sr.**: I met him for the first time at the AAU Nationals last summer.

**Louis Davis**: I was there; no, wait, not at that one, it was the one held in 2021. I went to that one. I was there in attendance.

**GM Sterling Chase Sr.**: Yeah, we were there for 2022, because after, I brought three athletes, and that was our first out-of-state competition. They did good. I'm proud of them. We're moving on!

**Louis Davis**: Were you to do this whole thing all over again to try and reconstitute a team representing the Air Force, what would you do differently; with the knowledge you have right now, what would you do differently?

**GM Sterling Chase Sr.**: The hardest factor that I had was because I was a low-ranking Black Belt. So, when I started with all of this, I was a 2nd Dan. So, here's this 2nd Dan talking to some 6th Dans on politics. They didn't want to talk to me. You've been around the martial arts arena, if you're not in that instructor 4, 5, and 6, and 7, 8, 9 Degree realm, they'll be like, "Okay, okay, go over there." So, nothing gets done and nobody's listening.

So, a couple of times I had to raise my voice and that's when, like Boatner and Harris, and even Jay Dunston, and they would tell the other Koreans, "Don't mess with him because he *will* go off on you."

I was the one that read the articles, the by-laws of the US Olympic Committee, I was the one that read the USTU by-laws. I talked it over with Professor Dong Ja Yang. He was the one who wrote them. He gave me his direct number in case I had any questions. All I was trying to do was try to understand this realm outside of the athletes and the competitors, because the political realm is totally different, and they're following rules of order, which was something new for me, too. So, I'm walking around with three volumes of Board recordings so I can make these meetings and come out with helpful information for our athletes.

So yeah, it was all a learning experience, and I was pretty hardcore on that, because as an NCO in the Air Force, you don't mess with my troops. You know what I'm saying? You don't mess with my troops, and every one of those Air Force athletes were like my troops.

And they were getting handed shady shit. I mean, I was getting calls for people stationed in other countries like Spain. I even got a call from a guy in the Philippines, wanted me to check on their Kukkiwon certifications and come to find out they had fake Kukkiwon certificates because I was in communication with Kukkiwon itself. We were running into some really difficult times where it was instructors that were selling fake Kukkiwon certificates for a very high price.

So, they were calling me, saying, "Hey, can you help me?" So, that's where I was at. I was the vocal point.

**Louis Davis**: Hmm. Someone else I heard of had similar power with the Kukkiwon. Are you familiar with Bobby Clayton?

**GM Sterling Chase Sr.**: Oh, yeah, I know Bobby pretty well. I haven't talked to him in years. I ran into him while I was in Korea one year.

**Louis Davis**: To my knowledge he is currently living in Korea right now.

**GM Sterling Chase Sr.**: Yeah, that's what I heard, too.

**Louis Davis**: ...between Korea and Maryland, that's what I've heard. Speaking of Grandmaster Clayton, I assume you've seen this man compete.

**GM Sterling Chase Sr.**: Oh, yeah, definitely. I saw him competing at the CISM in Korea.

**Louis Davis**: Okay, I am all ears, Sir. What was it like watching this man compete?

**GM Sterling Chase Sr.**: It was once, like, you know. I don't know if you've ever watched Patrice Remarck fight.

**Louis Davis**: I've trained under him.

**GM Sterling Chase Sr.**: Okay, yeah, see? I know. Well, Sumori Alpha is one of my good friends and he helped me with the political stuff. How deep did you want me to go?

**Louis Davis**: Hey, look, man, we got another twenty-five minutes. So, give me an idea of your background.

**GM Sterling Chase Sr.**: Lee Sang Cho, he's my uncle, in Taekwondo because my instructor gave him his first dojang in Binghamton, New York.

So, John Holloway and all of them, I've known them for well over forty years now. Yeah, I know all of them, yeah. We've had Soju and all that kind of stuff but they kind of punked out because they don't drink it anymore. But I'm a true believer!

**Louis Davis**: I learned my lessons drinking that stuff.

**GM Sterling Chase Sr.**: Oh, yeah, I still drink it. In fact, I talked to GM Sumori Alpha about three weeks ago. He called me.

**Louis Davis**: Sorry for interrupting. Grandmaster Alpha? He is Patrice's instructor, right? His coach? I met him at the Taekwondo Hall of Fame back in August.

**GM Sterling Chase Sr.**: Oh, yeah. So, he just had hip surgery, and he's got to get his other hip done. He's so full. I mean, "I need to get them both at the same time." I'm like, "They ain't gonna let you do that, no matter how tough you are." But yeah, so you know, like Leon Preston.

**Louis Davis**: Oh, no! GM Preston?

**GM Sterling Chase Sr.**: I had to fight Leon for my 1st Dan. I've been around, I've been around.

**Louis Davis**: Grandmaster Preston once gave me valuable information.

**GM Sterling Chase Sr.**: Oh yeah, in fact I talked with him last summer. There was GM Wilson out of California, Han Won Lee, the little Finweight fighter. That was his instructor.

**Louis Davis**: Geez!

**GM Sterling Chase Sr.**: So, in fact, I talked to him. Last time I talked to him was right before Covid. So, I still try and stay in touch, and it's so funny because a lot of people say like, "Do you know this person?" And I'm like, yeah but I really knew their instructor.

**Louis Davis**: Yeah, Master Preston. When I first got stationed in Washington back in 2009, I had gone into a school called Yi Sport that he helped establish, and during one of my training sessions he was sitting there and he recognized me right off the bat, but I hadn't seen Master Preston since the 1999-Armed Forces Championships. I'd forgotten who he was. He had not forgotten me.

**GM Sterling Chase Sr.**: Oh, no he doesn't forget. In fact, we're supposed to be doing another conference call. They want to do it this year and this call is for all the brethren that were with the USTU. So, there was quite a few of us. And I've got contacts for most of them. In fact, the last one that I contacted was GM Partridge.

**Louis Davis**: Barry Partridge?

**GM Sterling Chase Sr.**: He's got, what, three schools?

**Louis Davis**: Winston-Salem, North Carolina. You're kidding?

**GM Sterling Chase Sr.:** Yeah, yeah. I talked to him right before Halloween because we were talking about Halloween stuff, and how I was dressing for a Halloween party.

**Louis Davis:** He trained me during my two years stationed at Fort Bragg.

**GM Sterling Chase Sr.:** There you go. Yeah.

**Louis Davis:** Oh, boy!

**GM Sterling Chase Sr.:** I know it brings up a lot of thoughts. Doesn't it?

**Louis Davis:** Good Lord! Get out of my head, man. Let me ask you this question. You mentioned the brethren.

Okay, so what kind of discrimination did you face in Taekwondo, because sometimes, from what I heard from the other people that were there during the early days of the Armed Forces program; a lot of times the US military fighters were getting pencil-whipped by some of the USTU referees.

**GM Sterling Chase Sr.:** Oh, yeah, what I mean? Yeah, I was one of the ones fighting for them behind the scenes.

**Louis Davis:** Do you remember a Mr. Paul Boltz?

**GM Sterling Chase Sr.:** It sounds familiar.

**Louis Davis:** Mr. Boltz was the director of the Army Taekwondo program at Fort Indiantown Gap.

**GM Sterling Chase Sr.:** I met him one time.

**Louis Davis:** The stories I'd heard about Mr. Boltz in his early days was that he was a bit of a firebrand, that he made no small qualms about getting up in your face, if it was necessary.

Did you attend the nationals in 1993? Here in Saint Paul, Minnesota, where I live?

**GM Sterling Chase Sr.:** No.

**Louis Davis:** There is a story circulating about him that year: the Army fighters, and I think a few of the Armed Forces players were being cheated during some of their matches, and of course, Mr. Boltz went straight up to the Master's table and gave all of them an earful. He warned them that if they continue to cheat the military athletes that he is going to take back every dollar that he sunk into the USTU.

**GM Sterling Chase Sr.:** Yeah, yeah. See, I was one of those that were, "Yeah, if you guys continue this, you'd better go get your attorneys because we'll pull out and you'll lose your group A status." I wasn't playing. Fuji Mora used to always run up and say "Sergeant Chase, please calm down, please calm down, why do we got lawyers over here?"

**Louis Davis:** How close did USTU come to having litigation brought against them by the military?

**GM Sterling Chase Sr.:** To my knowledge, none. I would bring stuff up because they would say, okay, well, you guys are not authorized to vote because you don't have the three schools with the twenty-five members, and that kind of stuff. I'm like, "You guys are full of crap. You can't do that." And I would tell them, I'm like, "You better go find your attorneys, go find your lawyers because you guys can't do that." See, I don't have time to sit there and bicker with them. I was married to a Korean and I studied Korean culture and language, and I could curse them out in their own language better than they could. I don't play and if you mess with my people, my military guys, you don't want to see me.

Duane and Jay Dunston, they were like "DAYYUUM, here we go, there we go." They would come back, and they'd be like, "You need a shot of Soju?" I'm like, "Yeah." Do they listen? Yeah. Finally.

**Louis Davis**: Was there any discrimination based on ethnicity?

**GM Sterling Chase Sr.**: You know, the ethnicity was, initially, and we all went through this. And it wasn't just in Taekwondo; it was all throughout martial arts because we had a lot of Asians that came over.

And because it's their native art sport, whatever you want to call it, they're the ones that ruled. So, even though a lot of them were really young coming over, they would throw that Asian card. "Well, this is from my mother so I know more than you."

So, they put themselves on pedestals and a lot of them will bring others and put *them* on pedestals because we didn't have the rank. See, we were all lower rank, and the ones that did have rank or could have had the rank, didn't keep up with the promotion stuff. I met a lot of guys that were from the Vietnam era that were promoted by the Korean Black Tigers because they were with the special forces. So, they come to the United States. They didn't continue to run a program, you know, to keep their Dans going up.

Did you know Bruce Twain from Vermont?

**Louis Davis**: That name sounds very familiar.

**GM Sterling Chase Sr.**: He was Chung Do Kwon and Bruce got his 1st Dan with the Black Tigers in Vietnam because he was with the 82nd Airborne Division.

So, and that we used to tell him, you know, we sit there at the meeting. These young guys will come by, and then everybody would too. So, I'm standing with Alpha, Bruce Twain, Professor Dong Ja Yang and John Holloway. You know, we're all standing here, "BS-ing" and stuff, as always, and then these young Koreans would come by and you know they'd be in their suits and stuff and they were like oh, that's Master such-and-such.

And of course, like, you know that's that. And then I even asked Bruce one time, I said, "You know, a lot of these guys are so young, you know, with their 5th, 6th and 7th, to be Black Belts."

You know, I was just curious how you feel about it, because I do his background and he was like, "You know, let them have it because I got my 1st Degree before their mother was born."

**Louis Davis**: Hmm...

**GM Sterling Chase Sr.**: But yeah, so I know I've met a lot of veterans.

You know, especially from the Vietnam era, that trained and got promoted through the Koreans while deployed in Vietnam. And that was a problem, because they didn't stay in with the ranking system to be promoted like the 8th Dan and 9th Dan and on, and stuff like that, you know. It's like, okay? Well, I got my name. I said, "What Dan are you?" The usual response, "Well, I stopped at 4th Dan."

So, the hierarchy was the big thing, and it wasn't until, like, within the last ten years that we finally got high-ranking non-Asians in there because they put their dues in.

You know, take somebody like me, I've been training in Taekwondo over fifty years you know. I'm only a 7th Dan. I got a call last week from Jidokwon; they want me to be there this year because they wanted to promote me to 8th Dan.

How long has it been? Gerard Robbins told me that he was the first black man to get 9th Dan. He and I talked a lot about it, you know. A lot of things have been going on because we know a lot of the same people. There were a lot of changes that came because those of us stuck with it.

They kept up with the Kukkiwon to maintain our ranks, to keep getting promoted and then we're promoting each other, and that to me is where the goodness of what true martial arts is. We're promoting those that have fulfilled not only just the minimum requirements, but they've gone above and beyond to support the way, because that's what "do" is (in Taekwondo); it is the way.

So, to me that is where we should be, hopefully. We'll keep moving in that direction. And we'll keep them in that direction.

**Louis Davis**: I believe that's what Gerard's vision was when he established the Taekwondo Hall of Fame.

Which brings me to my next question: When did you first learn about the Taekwondo Hall of Fame? You brought up Mr. Robbins' name, so it's safe to assume that you know about it.

**GM Sterling Chase Sr.**: I remember when he started it and all the chaos because the Koreans didn't want him to have the title or the name of Taekwondo Hall of Fame. They were trying to take that away from him, and the brother had some good people behind him. And you know, we were all rooting for him and when that came through, the Taekwondo Hall of Fame was his. They were told to leave him alone and we were all proud of him. And we still are, you know, because he's gone where a lot of us have never gone before.

**Louis Davis**: We've reached the end of our time. Again, it was an honor to speak with you.

**GM Sterling Chase Sr.**: I look forward to talking to you more down the line.

The Air Force Taekwondo Team (year unknown)
What is widely known is that James Arrington was succeeded by Brad Carter as head coach of the Air Force Team. Unfortunately, I am unable to determine what year this picture was taken. Nonetheless, Brad Carter is beyond deserving of recognition for his efforts to keep Air Force Taekwondo moving forward.

# CHAPTER 7
# THE ARMY:
## BE ALL THAT YOU CAN BE

Members of the '86 Fort Bragg Team L – R: Laboy, Green and Leo Oledan
Photo courtesy of Pedro Laboy

"Many sports will keep soldiers in top physical condition if the soldier maintains the sport as a serious business, but to be one of the best takes time and a lot of discipline. The task of being a soldier and an athlete is not an easy task. You must get up early in the morning and get ready for PT, maintain your job skills and perform any additional task that lies ahead. Soldiers are soldiers twenty-four hours a day for seven days a week.

"However, becoming a soldier and an athlete is twice as hard, sacrifices in both your personal and family time must be made in order to get into top physical condition.

"Lots of soldiers studied Taekwondo before us, but as far as I know, they never represented the Army as a team. It was both myself and my close friend Pedro Laboy who took the first steps to officially represent the Army. At that time, the Navy and the Air Force had established teams before we did.

"When I reflect back to 1984 and all the things we did to gain support from our chain of command, it began with us training two hours prior PT formation (6:00 am) in order to prepare for both the North Carolina State and the US National Taekwondo Championships.

"At that time there were many obstacles that we had to overcome, first of all absolutely no support from our company commander, then there was the unit's mission, details, field training exercises, weapons qualification ranges, physical fitness tests and so on, but we were determined to see this through.

"All we had at that time was a shared vision, a dream and in order to accomplish that dream, we needed to first prove to our leadership at Fort Bragg (and later to the Army) that by demonstrating the highest quality of soldiering and athleticism we could effectively represent them well in this sport.

"Due to our success in 1984, the following year at Fort Bragg, North Carolina, the Army Taekwondo Team was formed." *Written by Rafael Medina, featured on his former website sport.tkd.center.com*

After hearing about what my predecessors went through to have a team representing the Army in this sport, it gave me a sense of pride. I am honored to be a part of this rich and nearly forgotten history which continues to be written even to this day.

Who would have thought that Rafael's words would carry so much weight? Even as I work to accurately document their journey I am amazed by the effort, diligence, and overall determination of both Pedro and Rafael. I can personally relate to their struggle to simply be recognized and supported by their leadership and of course, by the Army.

During my early years as an up-and-coming fighter stationed in Germany, I remained in constant contact with Coach Medina by phone, seeking guidance on how to best manage my time to train myself while in Germany. I too had very little support from my unit and *plenty* of people within my immediate chain of command that wanted me to fail.

Using the example that both of these men set enabled me to walk in similar footsteps. Like these two men I too had to show my chain of command (12th Chemical Company) that I had the potential to become a world class athlete.

My quest to uncover the origins of the Army Taekwondo Team led me to Pickens, South Carolina. It was Reginald Perry who sparked my curiosity. I remember listening to how passionately he spoke about his time with the team, his reverence for not just his time with the team but his fellow teammates, some of whom I knew personally from my first year with the team, others I'd heard about from those who helped me prepare for my first team trials while stationed at Fort Hood, Texas.

I spent the next few years trying to piece things together. It wasn't until early 2007 when Coach

Medina asked me to attend an event that he himself was unable to attend. It was around that time that he'd established a website for his dojang with a section dedicated to the Armed Forces Taekwondo team.

By God's grace, I have been able to speak with Coach Medina and Master Pedro Laboy together via Zoom.

## Zoom Call Interview with Rafael Medina and Pedro Laboy

**Louis Davis**: Gentlemen, I want to thank you, for taking the opportunity to join us in the Zoom call. Did both of you all receive the questions that I emailed you?

**Pedro Laboy**: Yes.

**Rafael Medina**: Yes.

**Louis Davis**: Let's dive right into it. My first question is for Mr. LaBoy. What motivated you to begin competing in Taekwondo representing the Army?

**Pedro Laboy**: It was a long journey that began before I actually got into the service.

Actually, before I enlisted, I had three international competitions and I enlisted. I was looking for the way to continue to improve my proficiency, you know, sharpen my skills so I could continue to compete at the international level.

So, when I finished basic training, I was assigned to Korea. After getting that assignment, I was just like a kid in a candy store because Korea is the birthplace of Taekwondo, and I was stationed there.

So, while stationed there I made contact with a few people there. I got the opportunity to compete at the Army level of Indian games over the area where our names were known.

**Louis Davis**: Which base was that over there?

**Pedro Laboy**: Over there it was in the 2nd Infantry Division (2ID).

**Louis Davis**: Right. Which base was it within the 2nd Infantry Division? Was it Camp Casey? Camp Red Cloud (CRC)? Which base, specifically?

**Pedro Laboy**: I was stationed at Camp Gary Owen, which is further north, but I was able to train with the local instructor in the nearby village outside Gary Owen while I was processing into my unit, and that's how I began to make contacts. This actually helped me to get promoted to my 2nd degree black belt under Kukkiwon.

The local instructor allowed me to train alongside the local Koreans who also trained at the dojang off post; this was in addition to my training on post. The only times that I went to Camp Casey was when they had the Indian Games. Competing in those games helped me to make new contacts.

After I left Korea and returned to the States around late 1983, early 1984, I was assigned to Fort Bragg, North Carolina. After I got there, I began looking for people to train with. I was always searching for ways to maintain and improve my skill set.

It began as me simply training myself and then I began to meet people that had the same things in common as myself and this led me to Grandmaster Myung Mayes, a local instructor who had a school in Spring Lake Park outside of Fort Bragg.

That's how everything started initially, at Fort Bragg.

**Louis Davis**: Master Medina, I will ask you the same question: what motivated you to begin competing in Taekwondo representing the Army?

**Rafael Medina**: Well, the thing that made me decide to do it at the time was because I was coming from Karate, boxing, kickboxing, so when I got to Fort Bragg, I was introduced to Master Laboy and then he began training me. He taught me the rules to WTF competition; WTF as an organization was well organized.

This was what motivated me initially to begin training and competing in Taekwondo, and Master Laboy was the only one I knew that was teaching this version of Taekwondo when I came to Fort Bragg from Korea. While I was stationed in Korea, I was learning Kuk Sool Won, and when I came to Bragg, that's when I met Laboy through my friend Sergeant Felix Arroyo, and we became friends.

After speaking with me for less than ten minutes, Laboy saw something in me and knew that I was hungry to do something, so he invited me to become part of his group and I said sure. During that time, I remember him asking me what my belt rank was in ITF Taekwondo (which I also had training in when I was in Puerto Rico). I told him that I was a Red Belt with a black stripe.

He tells me to start with him as a red belt in WTF which I promptly said no to, and I said I'll start as a white belt. A week later he tells me that "You gotta put some color there," so I say to him what about green belt? He agreed; however, because I learned everything so fast, he then told me to wear the red belt.

Like I've said, it was the WTF rules and of course Bruce Lee that were my inspirations to begin training in martial arts.

**Louis Davis**: Bruce Lee, you said?

**Rafael Medina**: Yeah, I was 14, and that's, you know, that's really the way got I started training and competing in Taekwondo, the World Taekwondo Federation. Yeah. Because I've been in so many different martial arts, but this one caught my attention and kept me. So, I'm happy to be part of the family.

**Louis Davis**: So, this next question is more along the lines for both of you.

**Pedro Laboy**: Let me add something to that. But I mean, you know. Please. Yeah, he didn't. He wanted to mention it, but I mentioned that I respected his previous rank as an ITF red belt, and I wanted him to go from there to learn the system and everything involved and then afterwards I would promote him to black belt.

I was trying to be fair with him because he was a fast learner.

**Rafael Medina**: He was really humble when he came to me.

**Louis Davis**: These next two questions are more for both of you. Now, when you decided to establish a team at Fort Bragg, how did that come about? And how were you guys able to receive support from both MWR and your respective units?

**Rafael Medina**: Laboy, can you answer that question?

**Pedro Laboy**: It was interesting. Let me put this in chronological order. When I first got there to Fort Bragg in 1984, I started training myself, and I got together with a few people who were also interested in training and competing.

That's also how Rafael Medina got involved and things got to the point that we decided to put together a team. Initially it was an unofficial team; I think that we called our team "The Fort Bragg" Taekwondo team, because they didn't have any other team representing the post at that time.

Prior to that everyone else was on their own. There was this one lady who was in control of Olympic Taekwondo in that area named Master Mayes. At that time, she was actively training a few soldiers from Bragg at her dojang off post.

But we were the only ones training on post and then we also training with her off post, and we were gathering other people who were interested in being a part of what we were doing, and that's how everything got started.

In order to compete in the North Carolina State Championships, you were required to be a member of a club recognized by USTU and so we were a part of her club, but not as members of her school. She acted as a sponsor for our team because we were doing things on our own.

1985 was the first time our team went to the North Carolina State Championships, and we successfully qualified to compete at the upcoming USTU National Championships. There were about six of us and every single one of us placed within the top three places in their individual weight classes. Our success provided us with an opportunity to go to the Nationals.

**Louis Davis**: So, in order to go to the Nationals, you had to qualify by placing within the top four, am I correct?

**Pedro Laboy**: Yeah, because our success at the State Championships created a *lot* of noise; it became publicity for us and for Fort Bragg. That's how we got in touch with MWR and got support from them. We told them that we're from Fort Bragg and we needed support. We needed a place to train in order to prepare for the Nationals.

Myung Mayes wrote letters to our chain of command and MWR, informing them that we had qualified to compete at the USTU National Championships, and that's how things started for us.

**Louis Davis**: I vaguely remember seeing a newspaper article that Master Medina provided commemorating your performance at the North Carolina State Championships. Were there any members of the base newspaper following you guys at the tournament?

**Pedro Laboy**: Yes, it was the *Paraglide Newspaper*. That was the on-post newspaper, and they followed up with us because this was big news, because this was a more than just a group of people, it was a group of US Army soldiers, plus the State Championships were held in Fayetteville and Master Mayes was the president of the North Carolina State Taekwondo Association.

Because the NC State Championships were held in Fayetteville that year, that's how we were given the opportunity to compete representing Fort Bragg. The interesting part of all of this is, because of what happened (our success at NC State), that is why MWR began to support us.

**Louis Davis**: What type of support did you begin to receive from Fort Bragg MWR?

**Pedro Laboy**: We were given a van as a means of transportation, we were given a place at the post gym to train, and as part of our training in preparation for the National Championships, we often traveled to Myung Mayes' dojang. We also traveled to nearby invitational tournaments held in both North and South Carolina; an example is a tournament hosted by Master Marlon of South Carolina.

These tournaments and additional training helped us sharpen our skills before the big event. At that time the WTF style of sparring was still a fairly new system, and it wasn't widely known except for a very small few.

This lack of familiarity was also the reason why we were so successful, because we were using these new methods of training and sparring. Skill-wise, knowledge-wise, we were at the national level.

Our support came and they weren't exactly "orders."

**Louis Davis**: What kind of support did you get at the unit level, and how high up the chain of command was that support given?

**Pedro Laboy**: Well first, they wanted to see the letters qualifying us to compete, they wanted to see the *Paraglide* article, basically all the supporting documentation so they could support us. I guess that some of the funds came from Fort Bragg MWR.

**Louis Davis**: What other kinds of support did you guys receive? Was it TDY Per Diem? What else did they give you?

**Pedro Laboy**: Before we got support from our units, there were some questions that needed some answers. At the unit level, nobody really knew what Taekwondo was; it was all something new to them, there were concerns that we were simply trying to get out of our normal duties. Others were wondering if what we were doing is real or just another in a long line of "new sports."

That was really a tough battle, getting support from our units. We had to prove ourselves, we had to demonstrate how serious we were about competing. Once we had that support, we then proceeded to Connecticut. It was at the US National Championships in Connecticut that the famous black T-shirts with our branch of service printed on the back came to life. That was the first time we displayed those T-shirts, "Fort Bragg Taekwondo Team."

**Louis Davis**: Was it Fort Bragg or US Army? What was printed on the back of those T-shirts?

**Pedro Laboy**: It was Army because we're actually representing the Army. Even though we were from Fort Bragg, we were representing the Army. This took place in 1985 and only a few people knew about this, to include Chief Baldwin of the US Navy.

**Louis Davis**: Circling back to the North Carolina State Championships, you said that there were six people from Fort Bragg competing. What weight classes did everyone compete in?

**Pedro Laboy**: Well…Medina was…

**Rafael Medina**: I was a Bantamweight! Leo Oledan was also Bantamweight and Mark Green was a Featherweight. I remember this because we had to fight each other all the time.

**Pedro Laboy**: Yeah, that's true. I fought in the heavyweight division. There was another person but I can't remember right now.

**Louis Davis**: I wanted to ask about Leo Oledan and Mark Green. When did you first meet Mark Green?

**Pedro Laboy**: I first met him while training on post at Fort Bragg. He saw us at the gym and he liked what he saw and decided to join us. That's how we got Mark Green on the team.

**Louis Davis**: What year did Leo Oledan join the team and how long was he with you guys?

**Pedro Laboy**: He was "with" us but he was a student of Myung Mayes; he was a product of her training and not ours. He spent more time training at her school off post than he did with us. I think that was because he lived off post. We still considered him to be a part of the team because he was a Fort Bragg soldier like us.

**Louis Davis**: Yeah, that makes sense.

**Pedro Laboy**: So, when we went to the Nationals, we had a lot of eyes on us. Now the Navy guys were better organized than we were. They'd been competing as a team a few years prior to us getting involved. It was around that time that Chief Baldwin approached us.

**Louis Davis**: When and where did he approach you?

**Pedro Laboy**: He approached us in 1985 at the 11th Taekwondo National Championships in Hartford, Connecticut. At that time, they had the Navy Taekwondo Association under USTU. After meeting Chief Baldwin I became the representative of the Army Taekwondo Association. I was the point of contact for the Army Taekwondo Association.

**Louis Davis**: Chief Baldwin mentioned that the Navy Team and your group had hosted tournaments between his Navy Team and your team there at Fort Bragg. How did this happen?

**Pedro Laboy**: It was part of a promotion to make more people on post aware of Taekwondo as a sport. Every Friday they used to have boxing smokers at Callahan gym and other gyms on post and we decided to put on a friendship Taekwondo match between the Navy and us. The Navy would bring their team to Fort Bragg. The exhibition matches were held in conjunction with the boxing matches.

We did this a few times and this really helped us promote Taekwondo as a sport but also the Fort Bragg Taekwondo team. People were talking about us and there were more people talking about the Navy team. Because of this we received more support at the unit level and from everyone else.

**Louis Davis**: When you began to receive support from your units, what were your training hours? How much time did your units give you to train in preparation for competition?

**Pedro Laboy**: Normally we weren't able to train together until the afternoon because in the morning we had to do unit level PT (Physical Training), but some of us were allowed to conduct PT on our own and we worked on out Taekwondo skills during that time. Everyone had a different situation with their individual units when it came to training in the morning during PT.

My unit allowed me to train in the morning, the afternoon and after duty hours unless there were some things that I had to do: PT test, weapons qualification, and stuff like that. We usually met at Callahan gym or other specified training locations to train after 1800 (6:00 p.m.). But in the morning, it was a different story. Our training in the morning depended on our individual units.

**Louis Davis**: Master Medina, did your unit allow you the same amount of time to train?

**Rafael Medina**: Yes, after our unit commanders saw the articles in both the *Paraglide* and the *Fayetteville Times*, that's when they decided to give us more time to train in preparation for any upcoming Taekwondo competition.

Because every unit at Fort Bragg had a different command policy, their protocols varied from unit to unit and the mindset of my commander was also different from Laboy's commander. As a result, I was allowed to train in the morning before PT from 0500 to 0630 and again in the afternoon from 1300 to about 1430; something like that.

In the morning, we were still required to show up at our respective units and go to first formation for accountability and any additional information that our chain of command had to put out to everyone.

My unit had two formations per day. The accountability formation at 0630 and the end of the duty day formation before 1700. In the evening Laboy would run an evening training session usually around 1800, and I would train myself again after I returned home.

I trained so much that my wife asked me "Hey, when are you gonna stop training?" and I say to her "Whenever I win." You know, it was an interesting time back then, like I've said, my Company Commander and especially my First Sergeant, they were my primary source of support at the unit level.

**Louis Davis**: I think that you've already answered this question earlier, but the first time the Fort Bragg Team competed at the USTU National Championships was in Connecticut, correct?

**Pedro Laboy**: Yes, that's right, it was in Hartford, Connecticut in 1985.

**Louis Davis**: What kind of financial support did you guys receive from Fort Bragg MWR? You mentioned that you were given a van to use for training and competition purposes; what else did MWR give you in terms of support? What additional support did Fort Bragg MWR provide when it came to getting you to the Nationals in Hartford?

**Pedro Laboy**: Fort Bragg MWR reached out to their division at the DA (Department of the Army) level, and as a result there were official TDY orders and some plane tickets so that we could fly to Connecticut, and any additional expenses that we had, we had to submit a team voucher at the end of the trip.

A year later we were invited by Chief Baldwin to the Armed Forces Taekwondo Championships at Little Creek, Virginia. This was a competition to select a team that will represent the United States at the World Military Championships (CISM Games) in South Korea.

**Louis Davis**: That competition took place the following year?

**Pedro Laboy**: The Army Team, which was organized and directed by DA Sports, went to the trials under orders from DA in 1987, but we (The Fort Bragg Taekwondo team) were previously invited to participate in the fight-off at Little Creek, Virginia in 1986, and the CISM Games took place in 1987. It was in 1987, after we came back from CISM, that's where we were introduced to Bruce Harris.

In that same year, Coach Bruce Harris organized the first All-Army Taekwondo Team trial camp at Fort Indiantown Gap, Pennsylvania.

**Louis Davis**: What year was the team out of Little Creek selected and when did the entire Armed Forces Team go to Korea?

**Pedro Laboy**: CISM in Korea? That was '87.

**Rafael Medina**: Yeah, that was in 1987.

**Pedro Laboy**: The Team was selected at the end of 1986.

**Rafael Medina**: The Armed Forces Team went to Korea in October 1987.

**Pedro Laboy**: It's a little bit confusing, you see, we didn't know anything about Bruce Harris until we came to the team trials in Little Creek, Virginia. That's when Coach Baldwin mentioned something to us about Bruce before we arrived at Little Creek. I remember him talking about having an Army guy there from Virginia working with him and they were trying to make this whole thing work.

Coach Baldwin introduced us to Bruce Harris and the idea was to connect Army guys with other Army guys so that we could network with one another.

**Louis Davis**: According to Chief Baldwin, Bruce Harris was his assistant coach for the Armed Forces Team. Is this true?

**Pedro Laboy**: Yes, it was true. Bruce Harris was the assistant coach and Chief Baldwin was the head coach.

**Louis Davis**: According to both Chief Baldwin and Master Medina, the tournament at Little Creek was a round-robin style tournament, so how many fights did each of you have that day?

**Rafael Medina**: Too many! Laboy had to fight for three days! I only had one day of fighting. The heavyweight bracket had a lot of competitors. I remember Laboy went through a *lot* of pain.

**Pedro Laboy**: There was a system we used called round-robin; the idea was that you had to fight each person in your bracket more than once.

**Rafael Medina**: Let me put it this way: say you've got ten guys in your division, that means that everyone's gotta fight nine times! *That's* why it's called round-robin. In Laboy's division he had almost twenty competitors, each from different styles of martial arts. You had Karate, Kung Fu, Taekwondo, you name it, there were so many different styles!

So, in a way the fighting was a mess but one thing that I know for sure was, everyone in the heavyweight division wanted to beat Laboy. Everyone was talking about Laboy, everyone wanted to beat him. He was the man to beat in that division. Everybody tried really hard to beat him, until he got injured.

**Pedro Laboy**: Yeah, my shins got a little swollen. A lot of the people came from traditional martial arts backgrounds and that type of thing. Sure, I also came from a traditional style, but I also came from a different style which was the Olympic style with the chest-guards. Their style was a little bit awkward and for some reason it was easy for me to work sidesteps, different kinds of footwork and counterattacks. So, it was easy for me to sack them. I wasn't gonna make it easier for them. A few guys tried to challenge me psychologically, you know, try to get in my head.

The people who challenged me the most were the guys that I fought really hard with, I mean, down to the wire. I was amazed at how everything fell into place during my matches.

**Louis Davis**: How many Army soldiers were there on the team?

**Pedro Laboy**: I think that there were about four or five of us, I mean, we had a lot. And then there were about three or four Navy guys and there was like one Air Force and two Marines, something like that. See, we had the team, and they took one alternate in certain weight classes.

**Louis Davis**: I see...my next question is: Once the team was selected, how long did you guys train in preparation for the competition?

**Rafael Medina**: Almost a month, right?

**Pedro Laboy**: Yeah, I think it was like a month to a month and a half at Little Creek, Virginia. To be honest, the preparation was what we needed at that time, and I have much love and respect for both Coach Baldwin and Coach Harris, but they wanted to train us in some kind of system that was simply outdated, and I was fighting with up-to-date techniques and footwork. That's just my honest opinion.

**Rafael Medina**: The techniques were more traditional techniques as opposed to the Olympic style sparring that was currently being used. That was their mentality. I could be wrong, but that's what everybody on the team was talking about.

**Louis Davis**: Once you guys arrived in Korea, what was the competition like? Tell me about the tournament. I saw a picture from the opening ceremony; what was it like standing in that environment? What were your thoughts during that moment?

**Rafael Medina**: To me? I was proud to be able to represent the United States, one of the biggest nations in the world. Seeing everyone standing at attention, lined up professionally, it wasn't just about me, it was us as the US Armed Forces representing the United States of America, representing our country.

Not only that but being the first team recognized by the Armed Forces Sports department and DOD. Prior to that, according to a chat I had with Chief Baldwin, there were no official teams representing anyone in this sport, people were told to go compete or one of those "Hey You" assignments.

I think that Laboy feels the same way as well as the other athletes involved. This was a big deal because it's not just one branch of service, but all branches of service were represented that day.

**Louis Davis**: Speaking of accomplishments, I understand that you were the first US Army soldier as well as the first Puerto Rican to medal at this event. Is that true?

**Rafael Medina**: Yes, but that's thanks to Laboy, I didn't do this by myself. He believed in me, and he placed a lot of emphasis on me, and even though he wasn't coaching me at CISM, I still trusted his words and his guidance because we were constantly talking about different things that really made the difference.

I don't have the words to express my gratitude for all of the things that I was able to accomplish because of him, because he helped me to take my first steps, he helped me to get where I am right now so I want to take this time to thank him for giving me the opportunity to become a part of something that's become larger than life.

I was the first DA- and DOD-recognized US Army soldier and the first Puerto Rican to medal at CISM; nobody can take that away from me. When I think about it, I'm like "Wow, I'm proud to be Puerto Rican!"

**Pedro Laboy**: I'm very proud of him for all that he has accomplished, not just as an Army soldier and athlete, but as a Puerto Rican.

During Mr. Laboy's reply he encountered an issue with his iPhone, and we lost him temporarily.

**Rafael Medina**: Before he comes back, I wanted to add something. I received a letter from the CISM president acknowledging my being the first US soldier to participate as an athlete, a coach, and an International Referee; this was out of, I think, 100 or 200 some-odd countries.

**Louis Davis**: It looks like we lost Mr. Laboy. Wait a second, it looks like he's back with us. It looks like he's still trying to iron things out.

**Rafael Medina**: Laboy has also done a lot as well. He took over quite a few things, he went through different channels for us to accomplish what he did to get support for us. It's because of his efforts, that's how we were able to accomplish everything that we did together.

We did everything by the book because he had to really hustle to get the support that we needed at that time.

**Louis Davis**: I quite understand. Master Medina, I want to take some time to focus specifically on your journey because at a specific point, the two of you went your separate ways. I don't know when the All-Army Taekwondo Team was officially established, my guess would be somewhere between 1980 and 1989, and Fort Indiantown Gap became the home of All-Army Taekwondo.

In 1991, war came: Operation Desert Shield and Operation Desert Storm. So how did the war in the Persian Gulf affect your ability to continue pursuing your dream of Taekwondo competition?

**Rafael Medina**: Can you repeat the question? I couldn't hear you.

**Louis Davis**: In 1991, Operation Desert Shield and Operation Desert Storm happened. How did that war affect both of you and your ability to continue training and competing?

**Pedro Laboy**: Actually, the war changed everything. Things got hot because the mission, the Army's mission, came first, everything else came second. As for Taekwondo training, we had to make time because everything else was geared towards deploying to Iraq. It was a very difficult time, that's for sure.

**Rafael Medina**: Remember, Louis, we were athletes, but the mission comes first. You know this better than anyone. When it's time to do our job during a time of war there *is* no choice because we are *soldiers* first; we deploy to fight for American freedom so guess what, we had to put our Taekwondo goals to the side and "go to work," you know, fight for our freedom. It didn't matter that you had a family or family things to take care of, when you gotta go, you gotta go, the mission is first.

**Louis Davis**: Did the both of you maintain contact with one another during your deployment?

**Pedro Laboy**: No, not really. It wasn't until after everything was settled that we were able to find out where each of us had gone after we got back from the war.

**Rafael Medina**: You know, Louis, it's funny, we lost contact with each other for I don't know *how* many years and out of the clear blue, everyone started talking! We didn't know where Mark Green had gone to, we didn't know where anyone was really, to be honest with you. Then he finally joined our alumni group. Ron Berry often stated that I was the glue that held everyone together because I always kept everyone posted on the Armed Forces Taekwondo Alumni website, www.sport.tkd.center.com/ArmedForces. The main goal of the website was to let everyone know that we're here and I didn't want anybody to disappear.

That's when we started having our reunions. We held it three times here in Hinesville and once in Orlando, and the most recent was in Louisiana.

**Louis Davis**: I definitely remember all of them. Moving forward, Mr. Laboy, just to be clear, you were appointed as the president of the Army Taekwondo Association via Myung Mayes, am I correct? How did that happen?

**Pedro Laboy**: That happened because of Chief Baldwin. He's the one that put my name there on the memorandum and they passed the position on to Bruce Harris because I was more of a competitor than a coach, and Bruce Harris was in the coaching department. I wrote, what I taught at that time was based on what I was taught previously. I wasn't into coaching, that's why Bruce Harris covered down on the coaching side of the house.

**Louis Davis**: What changed for you guys as soldiers, as athletes? How did you guys manage to balance these facets of your lives while serving in the military? How did you manage to balance training, maintaining your competitive skill set and of course, maintaining your military career. And assuming that the two of you were also married men with families, during your time in the ranks, how did you balance all of these things, how were you able to maintain your marriage, how were you able to keep these things from interfering with the others?

**Rafael Medina**: Do you know what our main source of support was? It was our wives that were our main source of support because without them, realistically we wouldn't have been able to accomplish what we did back then and even now.

I appreciate my wife a lot. Even though the Army gives us "family time" we still have to manage that time, between missions, training, being a husband and a father, being a good friend and sacrificing so that we can train hard. It's not an easy thing to do and Laboy can tell you it was *not* easy.

**Pedro Laboy**: I agree with Medina. It wasn't easy. Our wives had a lot to do with our success as athletes and as soldiers because they saw what we were going through, they experienced our frustrations, and they always encouraged us to always put our best foot forward. Their support meant a lot to us and that's why we appreciate their support.

Sometimes they say things like "Your unit and the gym see you more than I do and the only time I see you is when you go to sleep!" That is part of the sacrifice we made to become the best in Taekwondo, but *none* of that would have been possible without the support of our wives.

**Louis Davis**: As the saying goes, "Behind every great man there stands a great woman." There's a couple of remaining questions that I have for you guys and then we'll wrap this up.

All things come to an end, so my question is, when did you decide to leave the sport behind to focus on your Army career? What was the deciding factor?

**Rafael Medina**: Myself? To be honest I never "left" my martial art and that way of life behind; I was always doing something, I've been an instructor, I've been a coach but now I'm an international referee.

Next month I have to go to Mexico for referee training, so I will continue to be active and I'm sure that I'm gonna remain active until the day I die.

**Louis Davis**: What about during your time in the Army?

**Rafael Medina**: While I was still in the Army, if there was nothing to do, I would do something. I would take thirty minutes to practice basic kicking.

**Louis Davis**: You also managed to get promoted in the Army ranks, right?

**Rafael Medina**: Oh yes, like I said before, the mission comes first. So, if you don't take care of your military career guess what? You'll be running behind, you're not gonna make rank. You've gotta go to the schools to get promoted. So, when I was in both BNCOC (Basic Non-Commissioned Officer's Course) and ANOC (Advance Non-Commissioned Officer's Course), I studied my courses, but I still made time to maintain my Taekwondo skill, but I did that on my own time.

Really, I continued to train unless I had to study for a hard exam at either course, then I'd place all of the focus and discipline from Taekwondo into studying for my tests. There really wasn't much of a choice and failure wasn't an option for me.

Laboy always used to tell me, "To truly succeed you must be willing to sacrifice your time; manage your time well." This is one of the things that he would always tell me.

**Louis Davis**: Time management?

**Rafael Medina**: Yeah, this is what makes Laboy a great leader, not just for Taekwondo but for the community as well.

**Louis Davis**: Mr. Laboy, when did you decide to walk away from Taekwondo and the military?

**Pedro Laboy**: What Master Medina just said sounds just like what I used to tell him between rounds during competition: there's no time to rest, when it's over it's over, you really have to push forward and be willing to sacrifice.

Now to answer your question, in my case, it was a series of different things. I'll always have the sport in my heart but when I got hurt when I went to Egypt, when I hurt my back and I couldn't do the things that I used to, I used to be able to train but that held me back only a little.

What really changed things for me was the death of my son. My son died in a car accident. He was also in Taekwondo, and he was going to compete in the Junior Olympics and all those international events, and I was coaching him. I was very proud of his success because as a 16-year-old kid he used to fight in the adult division to sharpen his skills.

So, I was very proud of him and unfortunately while driving the car he got into a car accident. This was a hard blow for me because aside from teaching my son, I was also teaching a group in Orlando. After my son passed away it became very difficult to go to a tournament to see my other students because it would bring back flashbacks of my son.

That's what caused me to walk away from Taekwondo. But it was the church that helped me get back on my feet. They helped me address that part of me that had lingering questions and doubts. I still love the sport. Now I have my grandson following in his father's footsteps.

**Louis Davis**: This next question is centered around you, Coach Medina. When did you first become the assistant coach of the All-Army Taekwondo Team? Why were you chosen by Coach Bobby Clayton for that role?

**Rafael Medina**: Both Bobby Clayton and I were stationed in Korea around 1995. During a visit to Bobby's home, I learned that Bruce Harris had resigned as the coach of the All-Army Taekwondo Team. Bobby read the resignation letter to me stating that he had been recommended by Bruce Harris to assume the role of the head coach for the Army Team moving forward.

After reading the letter to me he then said that he wanted me as his assistant coach, since we were good friends. I stated that I didn't know anything about coaching. He then reassured me not to worry and that he would teach me.

He asked me again if I was willing to accept the job. After taking a couple of deep breaths I accepted his offer; I told him that I'll try as long as he would teach me. So, the following year I began my education into the world of coaching, learning everything that I could, and I was soaking it all up like a sponge, writing, watching all kinds of videos and then the following year, I grew even stronger as a coach. This was all thanks to Bobby teaching me what I needed to know. I'm very proud of Coach Clayton, he taught me well.

I'll never forget what I learned from Master Laboy because you never forget who helped you get started, who put you into position at the beginning. You gotta remember your roots which is your basics, which is what I did then, and I still do today.

**Louis Davis**: In 1998 I understand that you became the head coach for a little while or were you still an assistant coach? What happened that year?

**Rafael Medina**: In 1998, both Coach Clayton and I were unable to assume our roles as coaching staff for the team trials that year because we didn't have the support from our respective unit commanders, so we decided to recommend Coach Michael Bennett to take our place as head coach of the Army Team.

The team trials had been relocated from Fort Indiantown Gap to Fort Carson, Colorado that year, but something happened during the time because a decision was made to return the trial camp back to Fort Indiantown Gap, and somehow Coach Bennett was blamed for moving the team back to Fort Indiantown Gap.

I arrived at Fort Carson on a Thursday and returned to "The Gap" on Friday. Somehow, during this whole thing (I don't know how), Coach Bennett was fired from the position as head coach. I assumed the role as head coach and Bongseok Kim became my assistant coach at that time.

**Louis Davis**: Just in time for the Armed Forces team trials?

**Rafael Medina**: Yup, and after the team was selected, we went to the CISM games which were held at Fort Hood, Texas. We had a *very* strong team that year; we won five medals and Elizabeth Evans won the best athlete of the year award as well. Overall, the team placed third with our medal count. It was a very good year for the team.

**Louis Davis**: When did you become the head coach for the World Class Athlete program?

**Rafael Medina**: They were trying to bring Coach Hyun Suk Lee on board as coach of WCAP, but he was still coaching the Chinese Taipei team at that time. So, in order to get him here to the States he needed the required documentation. I was recommended by Coach Clayton and five other athletes to assume the role as coach of WCAP. So, I assumed the role from December of 1999 to January of 2000, which was a very short period of time.

**Louis Davis**: I understand you and Coach Clayton retired from the Army in 2000. What have you been doing since retiring from the Army? It's obvious that you've been very active in Taekwondo. Can you expound on that a little bit?

**Rafael Medina**: The main thing for me was, do my twenty years' active duty and get out. I ended up doing twenty-two years thanks to Coach Clayton because he wanted me to retire the same year he would retire. Actually, he retired three months before I did *but* we retired the same year because that's something that he wanted for the both of us.

The plus side of retirement is that you get half of your pay while you were in the ranks, *and* you can get another job making more money on top of it! That's why I retired, that, and to take care of my family.

**Louis Davis**: What have you been doing since you retired? How have you stayed involved with Taekwondo?

**Rafael Medina**: I started teaching at the Liberty County Community Center; teaching about thirty or forty children for free for about ten years, then I received a call from the city of Hinesville offering to pay me for my services as an instructor.

They asked me, how much did I want for my services, and I said let's try $25 per student, so it was $25 per student twice a week which they paid me a percentage of. I was happy with this arrangement. It covered my transportation expenses, gas money. I don't know who in the city made this decision to pay me but for some reason they decided to do this.

After I stopped teaching, I decided to attend the international referee course hosted by World Taekwondo. People that I knew always suggested that I become a referee, to include Bruce Harris and John Holloway. Both would tell me time and again, "Hey you need to do this!" One day I decided to take the plunge and I've been a referee since 2019.

**Louis Davis**: I also believe that you've received recognition from the Taekwondo Hall of Fame, am I correct?

**Rafael Medina**: Yes, when I first learned about the Taekwondo Hall of Fame, I spoke with the founder of the organization a few times because he needed a representative of the US Armed Forces, and at that time I couldn't attend because my job wouldn't give me the time off.

So, I asked a few of my former teammates if they could go, but none of them were able to attend. Finally, I spoke to somebody named Sergeant Louis Davis, who then became the representative of the US Armed Forces.

After he stepped down, CW4 Bongseok Kim assumed his role as both representative of the US Armed Forces and as Technical Advisor for the Taekwondo Hall of Fame. I'm very proud of you for stepping up and being the first representative of the US Armed Forces at the Taekwondo Hall of Fame. You were able to recommend a few people and overall, we're all very proud of you for doing that.

**Louis Davis**: Thank you, Sir. Thank you for supporting me with this monumental task of reconstructing the history of the Armed Forces Team.

**SPORTS**

*Taekwondo a way of life for two soldiers...*

# The Paraglide

# Taekwondo
## Ancient mysteries create a way of life

SECTION AB
THURSDAY, AUGUST 15, 1985

**Story and photos by Dave McNally**
*Corps Public Affairs Office*

"All things in nature operate according to rules. Mankind is one with nature." This is one of the basic principles of training in the martial art of Taekwondo.

Most people have a driving force in life, something that keeps them going, a philosophy that they are totally dedicated to. For Spec. 4 Pedro F. Laboy, of Headquarters and Headquarters Battery, 3rd Battalion (Airborne), 4th Air Defense Artillery, third-degree black belt, Taekwondo, is that force.

Recently, Laboy and Sgt. Rafael A. Medina, of Company B, 426th Signal Battalion, 35th Signal Brigade, a first degree black belt, competed in the 11th United States Taekwondo Championship at Hartford, Conn. Representing the U.S. Army on the North Carolina team, Laboy and Medina placed in the top 15 out of 100 competitors.

Some 16 years ago in his homeland of Puerto Rico, Laboy happily received a birthday present from his father: enrollment in Taekwondo classes. "When I started out it was like a hobby, I just wanted to learn about self-defense," he said.

And it was eight years ago that Laboy, who works as an Army mechanic, took the challenge of instructing others in the art of Taekwondo. "When I got to Fort Bragg two years ago I began teaching five to 10 students in my backyard, I find it very rewarding."

It's been a long road, but now Laboy teaches 30 to 40 students a week at the Soldier's Entertainment Center on Butner Road. At the center, which he named the "Yun Pung Gym," Laboy teaches for two hours each Monday and Wednesday evening. On Saturdays he offers special training.

Receiving a letter of appointment as a Taekwondo instructor from the World Taekwondo Federation, Laboy has become a credible instructor at Fort Bragg.

In Taekwondo, like Karate, there are degrees of skill. The bottom being white belt, the apex being 10th degree black belt. "When I first started out, my goal was to get green belt. I never thought then that I would get this far," Laboy said. Green belt is half-way up the ladder.

To achieve the rank of black belt, it takes years of serious study and practice. To the student, the rank is a symbol of accomplishment. But it is stressed by Taekwondo leaders that it should only be a symbol. "Students should enjoy walking the path, rather than worrying too much about the destination," the Federation says.

Being assigned to Korea, the birth place of Taekwondo, was a dream come true for Laboy. For the year and a half he was there, he competed as part of the Army team. "The atmosphere there really helps out. A lot of G.I.'s are involved in the sport," he said.

Laboy's crowning moment, however, came in 1979 when at the 2nd World Invitational Taekwondo Championships in Taiwan he won himself a bronze medal. "It was one of the greatest moments in my life." 87 countries participated in the meet.

"My goal is to get an official Fort Bragg Taekwondo team together, and maybe someday an Armed Forces team," Laboy said.

In the 1988 Olympics, for the first time ever, competitions will be held in Taekwondo. Twenty-nine year old Rafael Medina, Laboy's assistant instructor can already taste the fever of that future tournament. "Ultimately, I want to make the Army team and compete in the games," Medina said.

At 14, Medina who is a generator mechanic decided he wanted to learn how to defend himself. Much to his father's chagrin, Medina began to practice Karate. "I had to hide from my father when I practiced," he recalled.

Bouncing back and forth from Karate to Taekwondo, six months ago Medina dedicated himself totally to Taekwondo. Since then he has earned a first degree black belt.

Also previously assigned to Korea, Medina felt the land helped him out with his skill. "Some units in Korea let you get involved in the martial arts for PT (physical training)," Medina said.

"Taekwondo is more than a competitive sport," Laboy said. "It helps you out with your mind. It builds self-confidence."

According to the Federation, to simply classify Taekwondo as a "sport" denies its proud heritage of hundreds of centuries. "It's a way of life," Medina said.

"Taekwondo is knowing you don't have to fight, but if it comes down to it you can," Laboy said. "You have total control over your mind and body. You learn about people, about behavior, about life."

Laboy and his class will conduct a Taekwondo Tournament Aug. 24 at the Soldier's Entertainment Center.

An article in the Fort Bragg newspaper called the *Paraglide* about the Fort Bragg Taekwondo Team's success at the North Carolina State Championships

# Soldier
*From Page 1-B*

solve problems. Now I try to understand people, to understand the other person's position."

Although Medina says his expertise in the martial arts has gained him new respect that makes it unnecessary for him to fight with his fists, he has become a fighter in other ways. He came to the Army knowing no English and after 18 months was sent to Korea, where he had to struggle with a third language. But he returned with the Army's Achievement and Commendation medals.

Despite his language problems, his natural knack for mechanics earned him an E-4 rating in only 13 months.

"My friends said it was not fair because I could not speak English," Medina says in a still heavy accent. "But I was trying harder."

Medina says his hardships as a child gave him better understanding of others and have drawn him closer to his own children, a 5-year-old daughter and 3-year-old son.

But he says one of his most important lessons came from a martial arts instructor, whose philosophy was: "You can always win without having to fight."

Returning with newly gained self-confidence to his hometown, Medina said he found no one who wanted to fight with him because of his skills in karate and tae kwon do. And he's not looking for a fight but rather looks on martial arts as a hobby.

"People ask me what I would do if someone draws a knife," he says. "The first thing I would do is run. I am not a superman."

Medina has participated in the featherweight division in the Amateur Athletic Union tae kwon do championships in North Carolina and in national tae kwon do competition in Hartford, Conn. He hopes to get official sanction for a tae kwon do team at Ft. Bragg and to have an Army team compete in the 1988 Olympics.

"Tae kwon do is not just for learning self-defense, but it is based on a philosophy of discipline, self-control, meditation and a way of life," Medina says. "Accomplishing this has helped me very much in having self pride and being able to meet new people and appreciating the arts of life."

Part 2 of the *Paraglide* article

Mayes School of Korean Karate (Taekwondo)
308 Hope Mills Road
Fayetteville, North Carolina 28304
May 7, 1985

RE: Pedro P. Laboy (AAU NC Taekwondo Champion Heavy Weight Division) HHB 3/4 ADA Ft. Bragg, NC.

To Whom It May Concern:

The above captioned student from my school has qualified for the National Taekwondo Championships in Hartford, Connecticutt. In order to do this he had to come in first, second or third in his division at the North Carolina State Championships on May 4, 1985. The National Championships will be taking place on June 6, 7 and 8 (Thursday, Friday and Saturday) of 1985. On June 5 and 9 (Wednesday and Sunday) we will be spending the days traveling to and from the tournament. Participants in this tournament will be coming from all over the country and also had to place first, second or third in their state championships. Winners at the National Championship will qualify to go to the World Championships this year representing the United States of American and the black belt winners will go on to a place on the 1988 Olympic Taekwondo team. I would truly appreciate it if you could help my student to be able to participate in this worthwhile activity. Not only would he be representing my school and this state with his performance, but also the military of which he is a part. He could also go on to represent this country in international competition.

If further information is needed, please do not hesitate to contact me at the above address or by telephone. My number is (919)424-7903.

Sincerely,

Master Myong Mayes
6th Degree Black Belt
Mayes School of Korean Karate (Taekwondo)
National U.S.T.U. Taekwondo Championships:
  Recognized by - United States Olympic Committee
  Sanctioned by - The U.S. Taekwondo Union, Inc.
        and The World Taekwondo Federation

MM:sp

The letter written by Grandmaster Myung Mayes requesting support for the Fort Bragg Taekwondo Team Forces
Photo courtesy of: Pedro Laboy

## 11th United States Taekwondo Championship

# Certificate of Participation

This award is presented to

PEDRO LAMBOY-PEREZ

in recognition of participation in
the 11th United States Taekwondo Championship,
at Trinity College in Hartford, Connecticut

Presented in Connecticut
on this 8th day of June, 1985.

*Moo Yong Lee*
Moo Yong Lee
President

 **The United States Taekwondo Union, Inc.**
National Sports Governing Body for Taekwondo
U.S. Sole Representative Organization to the World Federation and Kukkiwon
Group A Member of the United States Olympic Committee

The Certificate of Participation of the 1985
National Championships in Connecticut
Photo courtesy of Pedro Laboy

Mark Green in action at Hartford, Connecticut
Photos courtesy of Pedro Laboy

March 3, 1986

To whom it may concern :

Pedro Laboy is a member of the U.S. Army , Fort Bragg Tae Kwon Do Team . On February 22 , 1986 he participated in the North Carolina U.S.T.U. State Tae Kwon Do Championships which was sanctioned by the U.S. Tae Kwon Do Union , Inc. and the World Tae Kwon Do Federation . This Championship has also been recognized by the United States Olmpic Committee .

Pedro Laboy placed 1 st in the Heavy Weight Division Sparring/Forms in this competition , thereby qualifying him to represent North Carolina and the U.S. Army , Fort Bragg at the National Championship in Dayton , Ohio , April 3-5 , 1986 .

This is a request for your support and sponsorship of Pedro Laboy , as well as the Fort Bragg Tae Kwon Do Team for this opportunity to represent the U.S. Army , Fort Bragg at the National Championship . This championship is recognized by the U.S. Olympic Committee . The 1st - 5th place winners at this event will go on to the World Cup event , with opportunities to attend the Olympic Training Camp and the World Sports Festival . They would also attend the Pan - American Games .

As you can see , this is an honor and an Exciting opportunity for Pedro Laboy , as well as the U.S. Army , Fort Bragg Tae Kwon Do Team . Your support is greatly appreciated .

VP *(signature)*

USTU Nationals, Dayton, Ohio

Results of the U.S. Army, Tae Kwon Do Team at the 2nd N.C. USTU Tae Kwon Do Championships held on February 22nd, 1986 at Fayetteville State University. Gold & Silver Medalists qualify to represent North Carolina at the National Championships.

Qualified:

| BLACK BELT DIVISION | SPARRING | |
|---|---|---|
| L. Oledan | Fly 110-118 lbs. | Gold Medal |
| R. Medina | Bantam 119-127 lbs. | Gold Medal |
| M. Green | Feather 128-142 lbs. | Gold Medal |
| B. Sparks | Middle 168-182 lbs. | Silver Medal |
| D. Marion | Middle 168-182 lbs. | Gold Medal |
| K. Chojnacki | Heavy 182 & Over | Silver Medal |
| P. Laboy | Heavy 182 & Over | Gold Medal |

| BLACK BELT DIVISION | FORMS | |
|---|---|---|
| P. Manning | Mens Division | Silver Medal |

| RED BELT DIVISION | SPARRING | |
|---|---|---|
| D. Howard | Middle 168-182 lbs. | Silver Medal |

**EVENT:** U.S. USTU TAE KWON DO CHAMPIOSHIP
**SANCTIONED BY:** The United States Tae Kwon Do Union
**RECOGNIZED BY:** The U.S. Olympic Committee
**DATE:** April 3-5, 1986
**PLACE:** Dayton Convention Center, Dayton, OH

MASTER MYONG S. MAYES (NAM KUNG)
MAYES SCHOOL OF TAE KWON DO
TOURNAMENT DIRECTOR
VICE-PRESIDENT N.C.-USTU

The following Tae Kwon Do competitors will participate in the National Tae Kwon Do Championships to be held in Dayto Ohio, 3-5 Apr 86

| NAME | RANK | SSN | UNIT |
|---|---|---|---|
| MEDINA, RAFAEL A | SGT | | HQ Btry 3/68 ADA, Dragon Bde |
| GREEN, MARK | SP4 | | 618th Engr Bn 82d Abn Div |
| MARION, DAVID C | SGT | | Svc Btry 5/8 FA XVIII Abn C |
| LABOY, PEREZ, PEDRO, F. | SP4 | | B Btry 3/4th ADA, 82d Abn B |

USTU Nationals, Dayton, Ohio

Photo courtesy of Pedro Laboy

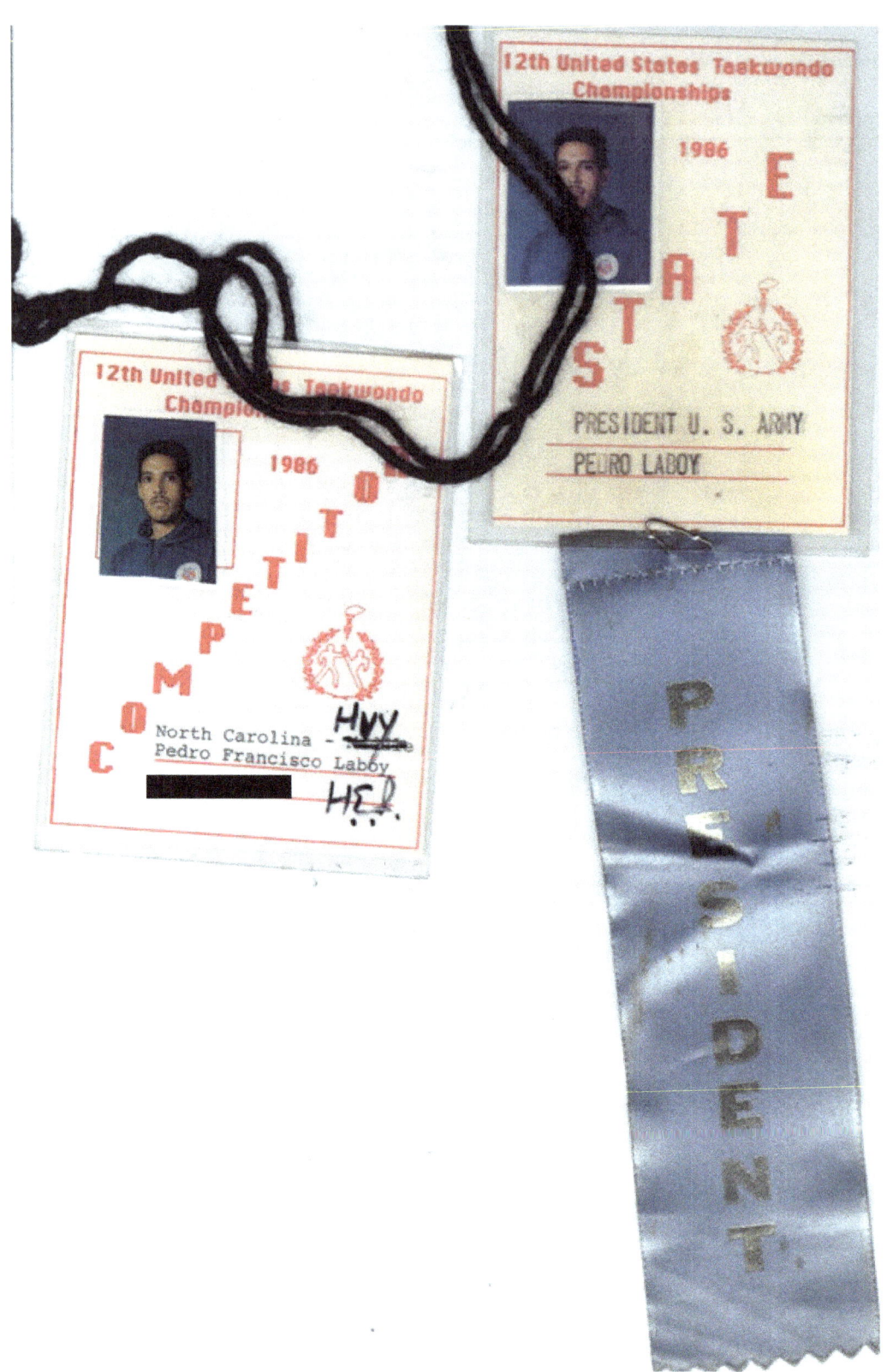

Photo courtesy of Pedro Laboy

Pedro Laboy in action at Dayton, Ohio
Photos courtesy of Pedro Laboy

# CHAPTER 8
# THE ROAD TO CISM

US Armed Forces Taekwondo Team CISM (1987)
Photo courtesy of Retired CPO William D. Baldwin October 25, 1987

After conducting three successful Zoom calls with these legendary pioneers of the Armed Forces Taekwondo Team, I decided to take a field trip to Mayfield, Kentucky to visit Chief Baldwin, who was gracious enough to allow me to stay with him during my visit.

During our many discussions, he spoke passionately about organizing the Armed Forces Taekwondo Team and the trial camp he successfully held at the Rockwell Hall Gymnasium on the Naval Amphibious base in Little Creek, Virginia.

Chief Baldwin stated that the camp itself was nothing short of brutal, and that a message was sent out to all intramural sports directors seeking potential applicants for this trial camp, informing the applicants that they must be in top physical condition *prior* to their arrival. The message further warned that this trial camp was not the place to "get into shape" and lastly, this competition was for those who are black belt level only.

Quite a few people applied from all four branches of the US Armed Forces with backgrounds in several known martial arts disciplines; under the name Taekwondo alone there were representatives of several different Kwons (schools), there were practitioners of various forms of Japanese Karate, and to my knowledge, at least one practitioner of Chinese Kung Fu.

On October 3, 1987, the first official Armed Forces Taekwondo team trials were held; the site that was selected to host the trial camp was a Naval Amphibious base located in Little Creek, Virginia.

Rockwell Hall Gymnasium, which served as the home of Navy Taekwondo, also became the home of the first Armed Forces Taekwondo Team.

The selectees came from the four corners of the globe, from bases within the United States, Hawaii, Germany, South Korea, and Australia. Make no mistake, the selection process was stringent and only the best of the best would be selected. Those who made the cut received a selection notice via their local sports office, followed shortly by a set of movement orders (TDY) enabling them to travel to Virginia.

The event would be a two-week trial camp where the athletes would endure a grueling training regimen consisting of four major practices per day, with Chief Baldwin taking video footage of each of the candidates and using the footage to evaluate each of the athletes' potential as competitors. He also made this footage available to the athletes themselves.

He brought in Master Sergeant Class Bruce Harris as his assistant coach. Bruce would be responsible for some of the administrative work such as conducting initial intake of the competitors and briefing them on the expectations of the camp, ensuring that they all understood the task at hand and the challenges that they would soon face, and the physical conditioning the athletes would endure, while Chief Baldwin focused on their fighting techniques and ensured that they were familiar with the current World Taekwondo Federation competition rules.

Each of the athletes would attend a referee seminar conducted by both Chief Baldwin and Sergeant Harris. This seminar ensured that the athletes were well informed and understood the calls a referee might make during competition. At the end of the seminar each of them was certified as referee under the US Taekwondo Union.

On Monday, October 5, 1987, those service members learned very quickly that Chief Baldwin's training regimen was grueling. The regimen consisted of a timed training circuit intended to simulate what the competitors would experience in Korea.

The circuit consisted of skipping rope, timed kicking drills and timed heavy bag drills, and eventually rotating out to a rest period. Initially he took it easy on them, having them spend 30 seconds per station.

When he turned up the heat, the athletes were required to sustain a high level of intensity at each station for a total of three minutes of intensity per station. Chief Baldwin had staged a few steel garbage cans for those who threw up during the training regimen. The rope drills were performed as a group; however, the remaining circuit drills were done individually.

To ease the pain of the regimen, Chief Baldwin had one-gallon tubs of Bengay positioned in the training area to address the amount of muscle soreness that the athletes would experience as a result of the training regimen, which often left the gym smelling of Bengay.

Each of the service-member representatives of the four main branches of the US Armed Forces vying for a place on the team began to bond as a team; however, true to military history, the camp had its fair share of branch as well as individual rivalry. As the time of the fight-off drew close, it seemed that all eyes were on a certain Army heavyweight named Pedro Laboy.

The matches took place over a three-day period and were to be conducted round-robin style, meaning you had to fight (and win) every match. Rafael Medina from the US Army easily won as a bantamweight, Timothy Hightower of the US Navy successfully captured the welterweight category, but all eyes were on the heavyweight division which had to include Pedro Laboy, a little over 20 competitors.

According to Rafael Medina, due to the sheer number of competitors in that weight division, Pedro fought the entire three-day duration to secure a place on the Armed Forces Team. What isn't well known is that an extremely talented up-and-coming lightweight competitor named Bobby Clayton, representing the US Army, was among the group of hopefuls competing for a place on the team slated to compete in Korea.

According to an article featured in a magazine called the *USTU Taekwondo Journal*, Mr. Clayton was an athlete that would have effortlessly earned himself a place on the team. As one competitor put it, Bobby Clayton was a human fighting machine. However, as fate would have it, Mr. Clayton was unable to compete due to an unforeseen hip injury and was forced to withdraw from the competition.

During the round-robin competition, the level of skill among the competitors within each weight class became apparent, and glaringly obvious. A small handful of service-member competitors was selected to become a part of the first Armed Forces Taekwondo team.

**Those service members were as follows:**

Representing the US Air Force in the Flyweight Division, Bennie Bryant

Representing the US Navy in the Featherweight Division, Mike Delgado

Representing the US Army in the Lightweight Division, Charlie Smith

Representing the US Navy in the Welterweight Division, Timothy Hightower

Representing the US Navy as an alternate in the Welterweight Division, Eric Hampton

Representing the US Navy in the Middleweight Division, Aaron "Ski" Cudnohufsky

Representing the US Navy as an alternate in the Middleweight Division, Melvin Boatner

Representing the US Army in the Heavyweight Division, Pedro Laboy

1987 US Armed Forces Team during CISM Opening Ceremony
Photo courtesy of Retired CPO William D. Baldwin October 25, 1987

# CHAPTER 9
# FORT INDIANTOWN GAP:
## THE RISE OF THE ALL-ARMY TEAM

The year 1988 was a pivotal moment not only for the sport of Taekwondo, but for the US military's involvement with the sport overall. After a successful competition at the 1987 World Military Championships in Korea, Master Sergeant Bruce Harris took the initiative to garner further support for the US Army. While the details of what took place remain unclear, what is well known is that Fort Indiantown Gap, Pennsylvania, became the new home of the All-Army Taekwondo Team, newly recognized by the Department of the Army Sports (DA Sports).

While researching the origins of not only the Army but also the Armed Forces Taekwondo teams, there's never been a discussion that I've engaged in with any of my predecessors in this sport without bringing up Mr. Paul J. Boltz. Paul initially didn't want another Asian "Chop-Socky" sport to support; however, Mr. Steve Brown and Mr. Phil Cota insisted that he give our team one calendar year to prove itself (Judo was already firmly established at that time).

Mr. Paul J. Boltz, simply "Mr. Boltz," was a sports specialist for All-Army Sports at Fort Indiantown Gap, Pennsylvania. To many people

Mr. Paul J. Boltz

who wore the logo of their respective branch of service and the Armed Forces logo, Paul J. Boltz was a man who was feared in some respects and well respected in others.

Well, the Taekwondo team did not disappoint. To my knowledge, the team under Master Sergeant Harris' direction demonstrated the highest levels of discipline, athleticism, and professionalism. Any time Mr. Boltz walked into the gym where the team was training, the team would stop, the command "CHARYOT KYONG

YE" was given by coach Harris, and the entire team, including coach Harris himself, would bow to Mr. Boltz.

This left a lasting impression on not only Mr. Boltz but also the gym staff and any military personnel who witnessed this expressed discipline firsthand.

This was one of the earliest examples of the "Soldier-Athlete" concept. This blended display of military and martial discipline quickly helped to gain further support from those who had doubts about the team.

As I work to write about the history of the Armed Forces Taekwondo Team, I would dishonor his memory had I not written about him. Although Paul is no longer with us, his hard work shall not go unnoticed or unrecognized!

What isn't well known is that the program was presented to Paul Boltz in or around 1988, and what began as a trial sport blossomed into a powerhouse of a Taekwondo team. An email from Mr. Boltz himself, in which he commends the success of the Armed Forces Taekwondo Team at the 2006 CISM Taekwondo Championships, also offers some insight on how our team and "The Program" found a home at Fort Indiantown Gap, and how he became our strongest supporter:

### Paul J Boltz: The Driving Force Behind Armed Forces Taekwondo at Fort Indiantown Gap

"Coaches Kim and Carter,

*Eighteen years ago when Mr. Phil Cota, then Chief of Army MWR, called and said, 'Paulie, I got a new program for you.' Knowing him well I responded, 'if you can't spell it I don't want it.' He said, 'I'll call you back.'*

*They had some sort of competition the year before at a Navy Base, but he wanted it to become part of the All-Army Sports Calendar - thus, along with All-Army Women's Basketball and Volleyball, Men's and Women's Softball and Team Handball, Taekwondo was baptized as an All-Army Sport at Fort Indiantown Gap.*

*I had already been in Sports and MWR Administration for thirty years, but as I sat in the bleachers I asked myself 'what is this all about— people out there kicking hell out of each other, taking a bow, thanking each other and asking for more.' I thought to myself if any of those people get mad at one another we'll have an investigation for manslaughter.*

*I called Mr. Cota and said, 'please get your butt up here because I really don't believe you know what you got yourself into;' well, he did and he came back year after year, and if you remember, for three consecutive years he held his little daughter as he spoke and presented awards. Taekwondo was now a proven commodity for Soldier Athletes.*

*We brought Canada and United Kingdom to Indiantown for Friendship Exchanges and were praised by General Martineau of Canada for the sportsmanship and discipline of our program.*

*As I glanced at the photos you shared with me, one could not help reminiscence about some*

*One Team. One Fight. One Family.*

*really proud moments you, the team, and your predecessors reflected what is good about All-Army Sports - you did your part to erase the 'ugly American' image, no matter the host country - you set a standard for others to follow both on and off the mat.*

*You were deserving of All-Army status; in fact, we tried numerous times to have Taekwondo recognized on the same par with Boxing, Wrestling and others as an Armed Forces Sport.*

*Taekwondo placed as many Soldiers on the medal platform as did any other of its combat sports in National and International competition, it had four CISM Champions and five National Champions and two Olympic Sports Festival Champions.*

*It had coaches like Pedro Laboy, Rafael 'Tony' Medina, Bobby Clayton and Bruce Harris, our first official team coach: its Women included Rachel Ridenour, Jada Monroe, Yelena Pisarenko and Joanie Guerrero.*

*The ranks of men were people like nationally and internationally respected Bobby Clayton, Reginald Perry, Jay Utter, Louis Davis and present day highly respected National and International medalists David Bartlett and Steven Ostrander, and of course recent medalists Jonathan Fennell and Jamie Toyota.*

*Our Chiefs of Mission included Congressional Medal winners, and now you are blessed. With probably the most knowledgeable and Internationally respected Chiefs of Mission for the past five CISM World Championships - Colonel Thomas A. Allmon.*

*When considering your opponents during the recent CISM Championship represented thirty-two National Delegations, many whose Team Rosters were comprised of their National Champion and Olympic Team members, your entire team made an incredible performance, add to that it was the National Sport of Host Country South Korea. And we are deeply appreciative of the contributions made by each of you.*

*I just want you to be assured I am deeply grateful for what you did for All Army Sports during my tenure; you were always a sharp and respected representative of our country, and I was proud to have been associated with you.*

*I am always pleased the gauntlet was passed to Ms. Claudia Berwager because she cares and encourages you to be appreciative to have worn the livery of the Gold and Black and Team USA.*

*Thank you for a great journey and pass along my congratulations to your 2006 edition of Team USA; they are smart looking "Ambassadors of USA Sports."*

*Paul*

Mr. Boltz with Coach Bobby Clayton

*One Team. One Fight. One Family.*

**Tribute to Paul J. Boltz and Claudia Berwager**

I dedicate this next section to the two All-Army and Armed Forces sports specialists whose tireless work kept us fed, clothed, paid, and they fought tooth and nail to ensure that we had everything that we needed to select and field the best team possible. The two seemed inseparable when it came to our team. Mr. Boltz seemed more like the team's father figure and Ms. Berwager, the team's mother figure.

In their own unique way, both of them will always hold a special place in the hearts of all of our Armed Forces Taekwondo alumni.

Mr. Boltz with Rafael Medina and Claudia Berwager at the Central Pennsylvania Championships, May 2016; Mr. Boltz with Rafael Medina at the same event

Coach Bongseok Kim with Claudia Berwager

1998-Armed Forces Taekwondo Team Abrahm's Field House

## 10th World Military Taekwondo Championships (CISM)
Date: November 19-21, 1998
Place: Fort Hood, USA

| Participating Nations | Male - 14 (Canada, Cyprus, Germany, Greece, Italy, Jordan, Korea, Lebanon, Netherlands, Peru, Russia, Saudi Arabia, USA, Vietnam)<br>Female - 6 (Canada, Germany, Greece, Italy, Russia, USA) |
|---|---|

| Weight | Rank | Male | | Female | |
|---|---|---|---|---|---|
| | | Name | Nationality | Name | Nationality |
| Fin | 1 | H. KIM | Korea | | |
| | 2 | N. Kiet | Vietnam | | |
| | 3 | G. Alshamrani | Saudi Arabia | | |
| | | | | | |
| Fly | 1 | D. KO | Korea | N. Kloske | Germany |
| | 2 | A. Seethaler | Germany | R. Huedes | Canada |
| | 3 | P. Cannizzo | Italy | | |
| | | | | | |
| Bantam | 1 | D. Kyu | Korea | D. Creti | Germany |
| | 2 | E. Denk | Germany | M. Karpathaki | Greece |
| | 3 | C. Irakleous | Cyprus | | |
| | | C. Lomas | Peru | | |
| Feather | 1 | J. CHO | Korea | **E. Evans** | **USA** |
| | 2 | A. Stozzo | Italy | L. Borodina | Russia |
| | 3 | M. Shdaifat | Jordan | | |
| | | **K. Jones** | **USA** | | |
| Light | 1 | M. Hatert | Germany | | |
| | 2 | **A. Roberts** | **USA** | | |
| | 3 | M. Van Heuman | Netherlands | | |
| | | P. Lam | Vietnam | | |
| Welter | 1 | **P. Nelson** | **USA** | M. Drosidou | Greece |
| | 2 | D. Betz | Germany | N. Lund | Canada |
| | 3 | A. Cutugno | Italy | | |
| | | **M. Remington** | **USA** | | |
| Middle | 1 | D. Hwang | Korea | | |
| | 2 | S. Mandy | Canada | | |
| | 3 | G. Al Ghanim | Saudi Arabia | | |
| | | **E. Laurin** | **USA** | | |
| Heavy | 1 | I. Aqil | Jordan | A. Girg | Germany |
| | 2 | M. Nitschke | Germany | I. Bykovo | Russia |
| | 3 | J. Sim | Korea | | |
| | | | | | |

Coach Rafael Medina during his first year as head coach of the Army and Armed Forces Taekwondo team respectively, had much to be proud of during the World Military Championship

1998 US Armed Forces Taekwondo Team in Dress Uniform

**The First Official All-Army Taekwondo Trial Camp**

The first official All-Army Taekwondo Trial Camp was held March 24, 1988. To many of the individuals that I've spoken with regarding this historic event, the camp was nothing short of brutal. Coach Bruce Harris continued the training regimen established by Navy Chief William Baldwin during the 1987 Armed Forces trial camp. This trial camp was fully funded by DA Sports. The soldiers attending this historic camp who successfully earned a place on the team would no longer have to bear out-of-pocket expenses associated with competing as an official representative of the US Army.

The first soldiers to attend the camp were James Barnes, Jay Debouse, Reginald Jones from Germany, Bobby Clayton, Jordan "JJ" Perry and Jada Monroe from South Korea, Curtis Gissander from the New York Recruiting Command, Barton Gonzales, Roger Jorah from Hawaii, Kermit Gonzalez-Irizarry, LaRoy Slaughter and Lynn McMillan from Fort Knox, Kentucky, Russell Hardy from Fort Carson, Colorado, Patrick Horrigan from Fort Ord, California, Pedro Laboy, Rafael Medina, Luis Torres and Steven Milosevich from Fort Bragg, North Carolina, Dwayne "Dewey" Lopp from Fort Campbell, Kentucky, Lynn Mariano from Fort Indiantown Gap, Pennsylvania, Roberto Montanez from White Sands Missile Range in New Mexico, Lawrence Powell from Fort Belvoir, Virginia, Sebastian Sciotti from the Pennsylvania Recruiting Command, and James Willis from Fort Sam Houston, Texas.

Coach Harris' primary focus during the inaugural trial camp was centered around the physical conditioning of the athletes, ensuring that they possessed the physical fitness necessary to compete in the sport. The athletes were reminded that they were to arrive at the trial camp in shape, ready to compete, and like the Armed Forces trial camp preceding it, this was *not* a camp to get into shape.

In conjunction with the competition training, Coach Harris also conducted a seminar to familiarize the athletes with the competition rules of both the US Taekwondo Union and the World Taekwondo Federation. Overall, the camp served as a screening period for choosing the best athletes possible to represent the Army at the upcoming national championships in Miami, Florida.

On March 30, 1988, the first All-Army Championships/Fight-offs were held at the Blue Mountain Sports Arena. The athletes were briefed on Coach Harris' selection criteria. Winning didn't always guarantee being selected nor did losing mean that you would be sent home. Once the team was selected, the team would then place greater emphasis on sparring, footwork, and more physical conditioning.

**Below is the official team roster of the 1988 All-Army Taekwondo Team:**

Lynn Mariano, Team Officer In Charge (OIC); Bobby Clayton, Team Captain, Lightweight; James Barnes III, Bantamweight; Rafael Medina, Bantamweight; Jay Debouse, Featherweight; Charlie Smith, Lightweight; Jordan"JJ" Perry, Welterweight; Luis Torres, Middleweight; Pedro Laboy, Heavyweight; Dwayne "Dewey" Lopp, Heavyweight; Jada Monroe, Women's Middleweight

The soldiers who didn't make the cut were still qualified to join the Army team at the 1988 National Championships but were required to attend at their own expense. It is unknown how many of these alternate soldiers attended the nationals in Miami; however, they deserve recognition as pioneers along with the official team.

**Those soldiers are as follows:**
Roberto Montanez, Flyweight
Roger Jorah, Flyweight
Lynn Mariano, Featherweight
James Willis, Lightweight
Barton Gonzales, Lightweight
Lawrence Powell, Welterweight
Kermit Gonzalez-Irizarry, Welterweight
Henry Block, Middleweight
Patrick Horrigan, Middleweight
Curtis Gissendaner, Middleweight
Russell Hardy, Heavyweight
Sebastian Sciotti, Heavyweight
Lynn McMillan, Women's Middleweight
Mary Michael, Women's Middleweight

After the trial camp, the US Army had enough soldiers to field nearly two complete teams which would make an impressive showing at the Nationals later that year. To ensure that there was a constant flow of both communication and information with other Army soldiers and their respective units, Coach Harris created the Army Taekwondo Association Newsletter, serving as an editorial to garner continued support Army-wide, and to attract new candidates for future All-Army trial camps.

Issue Number 1 was released in January 1988; this inaugural newsletter was distributed on a quarterly basis with the expressed intent of keeping the members of the Army Taekwondo Association, base sports directors, CONUS and OCONUS (Continental and Outside Continental United States), and soldiers who were actively competing in Taekwondo accurately informed of any and all upcoming events, changes within the national governing body, The US Taekwondo Union, and the World Taekwondo Federation.

These newsletters were packed with useful information, points of contact and of course, instructions on how to apply for any upcoming All-Army Team trial camp and the requirements necessary for selection to attend the upcoming trial camp.

As time went on, Coach Harris became a columnist for the *USTU Taekwondo Journal* and continued to educate the general public about the existence of the All-Army and Armed Forces Taekwondo program, and the newly established program which mirrored the Olympic Training Center's program for elite military athletes called the Army World Class Athlete Program, also known as WCAP.

He also encouraged any and all female athletes within the ranks of the US Armed Forces to apply for an opportunity to compete representing their respective branch of service. It was truly an exciting time, not only for Taekwondo but for the growing Armed Forces Taekwondo program.

**All-Army Strikes Gold at its First National Championships**

April 14 through April 16, the 14th annual USTU National Taekwondo Championships were held. The location was the James L. Knight Center in Miami, Florida. The first official All-Army Taekwondo Team not only fielded an impressively large group of soldier-athletes, but the team captain, Army Staff Sergeant Bobby Clayton, went on to become the Army's first national medalist securing for himself, the Army Team, and the US Armed Forces Taekwondo Program overall a place in history.

Although the US Navy Taekwondo Team's Eric Hampton bolstered the US Armed Forces presence at the USTU National Championships by capturing the bronze medal in the Men's Middleweight division, it seems that all eyes were on Staff Sergeant Bobby Clayton. According to details illustrated in Bruce Harris' April issue of the Army Taekwondo Association Newsletter, Staff Sergeant Clayton faced stiff competition as one of 102 competitors in the Men's Lightweight division.

Staff Sergeant Clayton won his first three matches quite easily; however, the fourth match proved to have a stiffer challenge. Staff Sergeant Clayton was afforded very little time to rest between his third and fourth matches, which affected his stamina levels, and his opponent in the fourth match was more rested. Despite this challenge Staff Sergeant Clayton was the victor, securing the bronze medal.

Staff Sergeant Clayton had his sights set upon capturing the gold medal, which became a strong test of his stamina by being allowed only a five-minute rest before the finals. All eyes were on Staff Sergeant Clayton squaring off against Indiana State's Garth Cooley, in anticipation of what would be noted in the Army Taekwondo Association Newsletter as, "One of the most electrifying matches to take place at the USTU National Championships."

The first round ended with neither player gaining a significant edge over the other, while there were some who felt that Staff Sergeant Clayton's technical superiority won the round. In the second round, Staff Sergeant Clayton was injured by Mr. Cooley after a failed attempt at a jump-kick, landing on Staff Sergeant Clayton's instep and severely spraining it.

Upon closer inspection by the tournament's medical staff, it was believed that Clayton's instep may have been fractured and they advised him to withdraw from the tournament. However, in a display of good old-fashioned Army intestinal fortitude, Staff Sergeant Clayton chose to continue the match.

Garth Cooley, aware of Clayton's injuries, attempted to use them to his advantage by pressing his attack, forcing Clayton into a more defensive posture, forcing him out of the ring, resulting in Clayton receiving two warnings. Staff Sergeant Clayton also began to experience leg cramping towards the end of round two but managed to hang on.

In round three, Staff Sergeant Clayton tapped into his inner strength, thus demonstrating the meaning of two of the tenets of Taekwondo: Perseverance and Indomitable Spirit. This enabled him to successfully counter Garth Cooley's attack strategy, resulting in forcing Mr. Cooley out of the ring twice himself. Staff Sergeant Clayton continued to press his advantage with a series of flurries, dazzling his opponent and exciting the crowd.

At the end of the match, it was obviously clear to the judges, referees, and the crowd that Staff Sergeant Bobby Clayton was the victor. His victory reflected great credit upon himself, his unit, and the US Army. This victory proved that the All-Army Taekwondo Team was here to stay.

**Team 1989: Doing More with Less**

Although the team had a strong first showing at the 1988 National Championships, they still faced the threat of not being funded by DA Sports to host and select a team for the 1989 season. However, with the help of Mr. Boltz, the team's strongest supporter, in conjunction with a few of his colleagues at DA Sports, the 1989 All-Army Team trial camp returned to Fort Indiantown Gap, Pennsylvania.

That year there was a total of 25 Soldier-athletes attending the team trials, many of whom were selected for the

All-Army Team the previous year while others attended the trial camp that same year.

The soldiers vying for slots on the men's team were Martin Calzadilla, Clifton Clark, Bobby Clayton, Barton Gonzalez, Russell Hardy, Patrick Horrigan, Gavin Hutchinson, Donald "Action" Jackson, Pedro Laboy, Dwayne "Dewey" Lopp, Rafael Medina, Steven Milosevich, Jordan "JJ" Perry, Reginald "Reggie" Perry, Lawrence Powell, Sebastian Sciotti Jr., LaRoy Slaughter, Charlie Smith, Luis Torres, and Steven Whittle. For the women's team, the competitors were Alana Conley, Michelle Franck, Lynn McMillan, and Jada Monroe.

The trial camp, although fully funded, encountered some unexpected changes. First, the time allotted to select, train and field this year's team would be done so on a truncated timetable. The athletes would arrive March 30 and the fight-offs for slots on the team were held 48 hours after their arrival at Fort Indiantown Gap. On April 2, life for those returning from duty overseas in locations such as Germany and Korea was difficult to adjust to, let alone for them to be prepared for their respective matches.

Team selections were made no later than that evening following the fight-offs, with non-selectees departing on April 3 and 4. It was during this truncated trial camp that both Mr. Boltz and Coach Harris stressed the importance of all future athletes arriving at the trial camp in tip-top physical condition and ready to compete.

With the new team selected, and very little time to prepare, Coach Harris still managed to put the team through a very rigorous training regime: an early morning run which Coach Harris jokingly called a "stroll"; that certainly wasn't what the athletes called it. Most of them referred to them as "death runs" which, according to former athletes such as Reginald Perry, Coach Harris ran with the team himself, barefoot!

Other strenuous exercises that were part of the regimen were goals of 500 sit-ups and 100 pushups per day. It's unknown if the team achieved this lofty objective, but according to Coach Harris, *everyone* either scored in the top 90 percent bracket of their Army Physical Fitness Test (APFT) or achieved the perfect score of 300 on the test.

True to Army Standards and traditions, Coach Harris sought to make improvements over the previous trial camp and raise the bar for the next team. He also saw a vast improvement in the athletes who made last year's team and in the athletes who were cut but were encouraged to return to the next trial camp.

Quite a few of the athletes improved in their running abilities and were able to keep pace with Coach Harris; he began referring to them as his "Stallions." There were so many stories of overall improvement of this year's Taekwondo Team over last year's team but there was one athlete named Sebastian Sciotti, who during the camp, successfully lost 35 pounds and was able to consistently keep that weight from returning. LaRoy Slaughter's sparring skills had vastly improved over his performance from the previous year.

Coach Harris continued to conduct seminars with the team, educating them on the current US Taekwondo Union and World Taekwondo Federation competition rules, ensuring that the team were up to date on any changes that they would encounter at the National Championships in Columbus, Ohio, April 13 – 15.

**The soldiers selected for the official team for 1989 were:**

**Men's Team**
Bantamweight: Donald "Action" Jackson
Lightweight and Team Captain: Bobby Clayton
Welterweight: Jordan "JJ" Perry
Welterweight: Lawrence Powell
Middleweight: Luis Torres
Heavyweight: Pedro Laboy
Heavyweight: Dwayne "Dewey" Lopp

**Women's Team**
Heavyweight: Alana Conley
Heavyweight: Jada Monroe
With Lynn Mariano serving as OIC (Officer in Charge)

1989 All-Army Taekwondo Team award ceremony post fight-offs

Official 1989 All-Army Taekwondo Team

Coach Bruce Harris addressing the athletes of the 1989 Trial Camp.
First formation of the 1989 All-Army Taekwondo Trial Camp

*One Team. One Fight. One Family.*

The "Next Generation" of the Women's All-Army Taekwondo Team

**Closing the Ranks. Continuing the Mission**

The following athletes who weren't selected were encouraged to participate at the Championships in Columbus, Ohio. Those athletes were Michael Baker, Martin Calzadilla, Clifton Clark, Barton Gonzalez, Russell Hardy, Patrick Horrigan, Gavin Hutchinson, Rafael Medina, Steven Milosevich, Reginald "Reggie" Perry, Sebastian Sciotti Jr., LaRoy Slaughter, Charlie Smith, Steven Whittle, Michelle Franck, and Lynn McMillan.

It remains unknow if any of these soldiers linked up with the official later that month, but they too deserve proper recognition. Reginald Perry returned the following year and won gold in his weight class (lightweight) and earned the opportunity to try out for the US National Team, while Rafael was chosen as an assistant coach to Bobby Clayton and later succeeded Bruce Harris as Head Coach of the All-Army Taekwondo Team.

**The Battle of Ohio: The Army Team Strikes Again!**

As the saying goes, "no rest for the weary." With the tournament just days away, the team intensified their efforts to prepare for the mission to Columbus, Ohio.

The training still consisted of a *long* morning run, followed by an unexpected but welcome addition to the team's training routine.

Bobby Clayton lent his knowledge of the modern training techniques and drills that were being used by the Koreans. He also brought in videotapes of competitions held in Korea, affording his teammates the opportunity to observe and study the high-level athletes of his home country. Bobby's act of selflessness and his demonstrated professionalism earned the praise of not only Coach Harris, but his teammates and Mr. Boltz.

The team, now fully trained and ready, left Fort Indiantown Gap and traveled to Columbus, Ohio to represent the Army at the 15th Annual USTU National Championships. Over the next three days, the All-Army Team would go head-to-head against the top male and female athletes representing all 50 states. Once again, the All-Army Taekwondo Team would rise to the challenge.

Unfortunately, the All-Army Team was not joined by representatives from the other branches of the US Armed Forces. Up first were both the Men's and Women's Finweight and Heavyweight divisions; up next were Lightweight, Bantamweight and Featherweight. Donald Jackson lived up to his nickname "Action," displaying incredible abilities as the Army's newest Bantamweight, with a mental focus as sharp as a razor, and easily secured a bronze medal, the first of three for the Army Team. Up next was Team Captain and Lightweight Bobby Clayton, who with surgical precision captured the Army's second bronze medal.

On the final day of competition, the Flyweight, Welterweight and Middleweight divisions rounded off the three days of the exciting Taekwondo sparring competition. On deck were Jordan Perry and Lawrence Powell competing in the Men's Welterweight division and Luis Torres competing in the Men's Middleweight division.

As the competition progressed, all eyes turned to the Men's Welterweight division and the Army Team's Jordan "JJ" Perry as he made short work of the competition, easily leading him to the semifinals where he secured the Army's third bronze medal.

The Army, with a truncated timetable, successfully selected, trained, fielded, and secured three Nationals medals, which qualified three Army Soldiers to attend the US Team trials. The Army had *much* to be proud of that day.

**The Mission to the 1989 US Team Trials**

The Army, after successfully qualifying three soldiers to compete for positions on the US National Team, prepared for their next mission, which was to successfully place their three heavy-hitters on the US National Team. Their hard-fought victories from the previous competition, however, came at a price as all three soldiers received injuries during the competition at Columbus, Ohio.

Army Taekwondo Team Captain Bobby Clayton, due to unknown complications, was unable to join his fellow teammates Donald Jackson and Jordan Perry at the US Team Trials at Colorado Springs. Perry experienced similar issues due to his unit's pending deployment to Panama, which offered him even less time to train in preparation for the trials.

The competition was conducted round-robin style requiring those who placed within the top four of their weight class to face each other, and the athlete with the best record was selected for Team USA. Donald Jackson re-injured himself during the preliminary matches and was forced to withdraw from competition. Jordan Perry fought valiantly and was able to secure an opportunity to compete at the upcoming Olympic Sports Festival to be held in Oklahoma later that year.

Although the All-Army Team missed the mark at this event, the performances of Bobby, Donald and Jordan earned them an article in the September 1989 issue of *Soldiers* magazine.

Another All-Army Taekwondo Team member, Pedro Laboy, took the bronze medal during the Pan-American Taekwondo Championships held in Lima, Peru. Overall, Coach Bruce Harris, with regards to this year's team and their overall performance, had much to be proud of as did the support staff at Fort Indiantown Gap.

Coach Harris then used this momentum to work towards creating a Taekwondo-based World Class Athlete Program (WCAP), which was originally established at Fort Myer, Virginia, and placed Bobby Clayton as its first resident athlete.

This new WCAP would offer All-Army Taekwondo athletes the opportunity to train full time for one calendar year to improve their Taekwondo skill set, with the overall goal of placing these athletes on the US National Team, and ultimately placing a soldier on the US Olympic team.

At the close of the decade, the overall success of the US Armed Forces Taekwondo Team at the 2nd World Military Championships (CISM) and within the US Taekwondo Union firmly established their presence within the sport—an accomplishment which was felt by all concerned during those early years.

If you ask anyone who faced these men and women on the mat, chances are that the common response would be that each of these military athletes was a force to be reckoned with. As I continue to write this story, I can only imagine what it must have been like during those early days. I will say that I am proud to have been part of the third generation of these amazing military athletes.

# Taekwondo Tourney

Stories and Photos by SFC Frank Cox

PRESIDENT George Bush gave a hearty welcome in a March 30 letter to everyone gathered at the 15th U.S. National Taekwondo Championships in Columbus, Ohio. "This martial art form is more than a demonstration of grace and physical strength," he said. "Its students not only become more physically and mentally fit but also grow in integrity, respect, self-discipline and motivation. Students of taekwondo learn valuable lessons . . . and inspire all of us who watch this wonderful sport."

It was inspiring to watch the Army taekwondo team perform during the mid-April event. Three soldiers won bronze medals and qualified to compete for berths on the U.S. taekwondo team. The bronze medalists were SSgt. Donald "Action" Jackson, Fort Carson, Colo.; Spec. Jordan "J.J." Perry, Fort Ord, Calif.; and team captain SSgt. Bobby Clayton, Eighth Army, South Korea. Clayton won a gold medal at the national meet last year.

Also on the Army team were 2nd Lt. Alana Conley, Eighth Army; SSgt. Jada Monroe, Fort Meade, Md.; Capt. Lawrence Powell, Fort Belvoir, Va.; Sgt. Dwayne Lopp and SSgt. Luis Torres, U.S. Army, Europe; SSgt. Gavin Hutchison, Fort Carson; and Spec. Pedro Laboy, Fort Bragg, N.C.

Laboy, a heavyweight, won a bronze medal in the Pan American taekwondo championships in December in Lima, Peru. He missed winning third place at the nationals by one-half point after his fourth bout. Hutchison, 24, a 138-pound featherweight, enlisted six years ago — nine years after he got into the martial arts. He's a five-time Pennsylvania state champ and, like other converts on the team, now prefers taekwondo. "It's getting tougher and tougher to compete at this level," he said, "but I'll definitely be back next year."

Recognized officially by the Army last year, taekwondo is a rising star in major commands. Soldiers on the Army team this year come from commands dotting the landscape of three continents. The members, all black belts, have been trained by masters who enforce a strict code of conduct that traces back 2,000 years to the time when the Korean martial art was developed as a combat discipline. Now, 100 million people in more than 100 countries have taken up this ancient art.

Amateur sport taekwondo focuses attention on the use of the feet. Hands are used mostly for blocking. Feet and hands may be used to kick and punch legal areas of the body. Punches or kicks to the groin are prohibited. Kicks to the head are legal and can be devastating.

Team members at the nationals expressed hope that taekwondo will become an official sport in future Olympic Games. Un Yong Kim, president of the 14-year-old World Taekwondo Federation, is working to that end.

"Taekwondo debuted in the 1988 Seoul Olympic Games as a demonstration sport. All the competitions were completed in front of sell-out crowds," he said. "It is now a matter of time before taekwondo will be adopted by the International Olympic Committee as an official sport."

The IOC, which formally recognized taekwondo in July 1980, had members present to observe the national meet.

MSgt. Bruce Harris, the Army team coach and a musician in the Army Band, "Pershing's Own," at Fort Myer, Va., hopes to put

*The team was well conditioned and trained.*

### J.J. and Grandpa Bear

HUNGARIAN-BORN Zoly Zachery Zamir and his wife flew from Houston, Texas to Columbus, Ohio, in April to surprise and support their grandson, Spec. Jordan "J.J." Perry, a top contender at the 15th U.S. National Taekwondo Championships. The Zamirs had raised Perry since he was a toddler, and all they'd "ever wanted was to have a good human being when he grew up."

Zamir's wish is coming true. Perry, now assigned to the 7th Battalion, 15th Field Artillery at Fort Ord, Calif., began taking the Korean style of karate at age 8 and earned his black belt at 13.

During the nationals, Perry went through four opponents on his way to the medal rounds. The last Army team member to compete, Perry was also the Army's last hope for a gold medal.

Between his fourth and fifth bouts — after a dozen three-minute rounds of total effort — Perry got a breather. As he ignored red welts and growing bruises, a protest was issued by his last opponent's coach. An outwardly calm Army team coach, MSgt. Bruce Harris, and two anxious grandparents stood nearby.

Ten long minutes passed before associate referee Robert Gross approached the group. "The ring judges questioned the corner judges, who had to explain why they scored as they did," he said. "The decision was upheld."

Cheering erupted from the Army team. "The judges ruled Perry the clearly superior fighter," Gross added.

But Perry was in pain. "It's my right knee and ankle," Perry said at the time, adding he thought he had some bone fragments floating around in the two joints. Perry faced a decision to either go for the gold at the nationals — and possibly aggravate his injuries — or "bow out." Urged by his coach, Perry excused himself. His move gave his injured body time to recover before the U.S. team trials in Colorado Springs, Colo., in May.

Zamir waited patiently while the bronze medal was hung on Perry's shoulders. Afterward, the burly grandfather gave his grandson a bearhug. Perry's grandmother beamed when her husband took off his glasses and twirled them victoriously. Their grandson, the soldier-martial artist, had become everything they wanted . . . and more.

*Spec. J.J. Perry plants a body-rocking, round-winning reverse kick during the nationals. During Army team trials at Fort Indiantown Gap, Pa., he ran up to 15 miles a day to build endurance.*

### Action Jackson

SGT. Donald Jackson began taking taekwondo 22 years ago, his first year as a teenager. A year later he switched to Chinese kung fu and stuck with it for 16 years. In 1983 he took on taekwondo, one of the most popular martial arts styles in the world. "Action" Jackson switched because his instructor, SSgt. Bobby Clayton, told him back in 1981 that taekwondo would come an Olympic sport.

Switching was difficult because he had to comply with strict Amateur Athletic Union and U.S. Taekwondo Union rules, which he's mastered. Last year Jackson was the AAU national champ in the featherweight division. Since then, he's dropped 10 pounds and one division, to bantamweight. At 130 pounds he packs leaner 35-year-old muscle and can move like an attacking cheetah.

At the 15th U.S. National Taekwondo Championships in April, the Fort Carson, Colo., medic used a well-developed strategy.

"I'm just taking a nice, easy pace because I don't want to burn out or show everything I have. I fight each opponent differently," he said. "I'll go into the first round easy to feel my opponent out. In the second I'll pick it up some, and in the third I'm going to 'cook.'"

Jackson took opponents apart round by round, bout after bout, at the nationals. Between rounds No. 2 and 3 in Jackson's third match, team captain Clayton whispered a few sage words in his ear. "I told him he was lifting his left leg and his opponent noticed, so instead of leading and faking with it, he should use it," Clayton said. Jackson obeyed and won a bronze medal.

In his final bout, a highly charged Jackson came out feinting, dodging, leading, looking for an opening and . . . suddenly it was over. He took a hard shot to his pelvis. In the third round he could barely raise his right leg, and his opponent took advantage. Afterward, Jackson was ordered to a hospital for X-rays.

"Just a bruise," he said later. It'd take a lot more than a few hard knocks to stop Action Jackson.

*Sgt. Donald "Action" Jackson rests between bouts. While fighting, though, he demolished one opponent after another. A medal round injury forced his withdrawal.*

Article in *Soldiers* Magazine, September 1989 Issue

## Doing It His Way

CSM Fredricke A. Clayton is more than proud of his adopted son, SSgt. Bobby Clayton, who last year won the national taekwondo championship in his weight division and then fought to within a point of making the Olympic team.

"Bobby has championship qualities across the board, in and out of the ring," said the command sergeant major of Walter Reed Army Medical Center in Washington, D.C.

The younger Clayton, stationed in Yongsan, South Korea, earned the right to compete for a berth on the 1989 national team in April when he placed third at the 15th U.S. National Taekwondo Championships.

Round No. 1 of his first bout proved Clayton a superior athlete. With feet moving almost too fast to see, the poker-faced lightweight remained on the offensive. He planted a reverse hook kick in the second round that could've staggered Mike Tyson. It floored Clayton's opponent, taking the fight out of him in the process.

Clayton beat his second opponent in a three-round decision and then, with only five minutes' rest, faced his third rival. A blurry reverse side kick sent his Virginia opponent reeling in the second round. The Virginian clearly went on the defensive. Another strong reverse kick to the civilian's chest was followed by a snapping round kick to the head. Fans in both camps winced with each stinging sound from the ring. Clayton's opponent couldn't stop the onslaught.

After eliminating two more rivals, Clayton got a breather and watched a bout, the winner of which he later fought and dismantled for the bronze. The silver and gold medals went to two fighters Clayton had beaten at the 1988 nationals, however.

During his final bout Clayton felt an old injury resurface. He bowed out because of the pain in his right hip, "the same injury that's been giving me trouble," he said.

Unfortunate. But he'd made it into the national arena.

One side of Clayton is as mild-mannered as Clark Kent, the other pure Tasmanian devil. "Bobby was born in 1963 in Munsan-ni, South Korea, only three miles from the Demilitarized Zone. He used to get beat up at the Korean school he went to — just because he's Amerasian," his dad said. "We adopted him when he was 6. He's practiced the martial arts since he was a child and is fluent in Korean and English. He's never been a bully.

"He played football at Seoul American High School, but that didn't really turn him on. Lifting weights and taekwondo are his lifestyle. He's always wanted to do it his way, and I support him.

"Bobby has been in the Army nine years and has never made less than 300 points on his PT test. His goal is to become an American champion. I think that says it all about the kind of soldier he is."

*SSgt. Bobby Clayton inspired taekwondo fans with his quiet, fluid style. An injury kept him out of the finals, from facing the two men he defeated last year for national gold.*

Army martial artists into the Olympic spotlight. "Last year was the first time an Army team went to the nationals, in Miami, where (Clayton) won the first-ever gold medal by any branch of the armed forces," he said.

Harris has a solid background in taekwondo and was recently certified to referee at the international level. He was selected as coach because of his experience and position as president of the Army Taekwondo Association, an appointment made by the president of the U.S. Taekwondo Union and the Army.

"Athletes on the Army team this year are better conditioned and trained in the style of fighting that leads to success at the international level," Harris said. "I encourage Army red through black belt martial artists to apply to the Department of the Army Sports Office — through their local commands — for a chance to attend the Army taekwondo camp next year."

When the 1990 nationals take place, it'll be a safe bet the medal count will go higher as soldiers kick their way toward international gold. □

Article written about Bobby Clayton in the September 1989 issue of *Soldiers* Magazine entitled "Doing It His Way"

The following pages contain the official results for both the 1988 and 1989 USTU National Championships.

## 14th National Taekwondo Championships

James L. Knight Center
Miami, FL
April 14-16, 1988

### Black belt gyo-roogi

**Fin**

Men
1st Juan Moreno (IL)
2nd Robert Leach (PA)
3rd Chuck Flayler (OH)
3rd Jeff Pinaroc (TX)

Women
1st Helen Yee (OH)
2nd Cheryl Kalanoc (IN)
3rd Susan Palmer (KY)
3rd Diana Radakovic (CA)

**Fly**

Men
1st Hyon K. Lee (CA)
2nd John Monroe (TX)
3rd Jeffrey Coffey (OK)
3rd Craig DeRosa (NY)

Women
1st Ginean Hatter (VA)
2nd Arabella Naguit (OH)
3rd Theresa Alvey (CA)
3rd Susan Park (AL)

**Bantam**

Men
1st Chris Spence (OH)
2nd Clay Barber (TX)
3rd Britney Combs (NV)
3rd Hee Chan Chung (FL)

Women
1st Nosrat Elyassi (MD)
2nd Susan Kim (NY)
3rd Jennifer Gray (OH)
3rd Heather Byron (LA)

**Feather**

Men
1st Raphael Park (WA)
2nd Tuoi Nguyen (WA)
3rd John Kim (NY)
3rd Dong S. Lee (TX)

Women
1st Kim Dotson (OH)
2nd Ayoka Brown (MD)
3rd Josephine Nyland (FL)
3rd Mai Nguyen (TX)

**Light**

Men
1st Bobby Clayton (AY)
2nd Garth Cooley (IN)
3rd D. Steve Shinn (MO)
3rd Kareem Ali Jabbar (IL)

Women
1st Carolyn Raimondi (CA)
2nd Dana Hee (CA)
3rd Gail Hinshaw-Wright (C
3rd Anne Louise Long (OH)

**Welter**

Men
1st Doug Baker (OH)
2nd Charles Thompson (MN)
3rd Mike Demkowski (IA)
3rd Eric Hampton (Navy)

Women
1st Michelle Pellegrini (CA)
2nd Diane Shaieb (CA)
3rd Susana Mirjanic (IL)
3rd Terri Bolduc (ME)

**Middle**

Men
1st Roland Ferrer (CA)
2nd Joon Yang (MD)
3rd Naeim Hasan (OR)
3rd Ed Shorter (MA)

Women
1st Lydia Zele (CA)
2nd Sharon Jewell (DC)
3rd Rhonda Juarez (TX)
3rd Diana Mason (MD)

**Heavy**

Men
1st Jimmy Kim (CA)
2nd Won S. Yang (MD)
3rd Scott Miranti (MT)
3rd Glenn Warren (IL)

Women
1st Kathy Wagner (CO)
2nd Gwen Teague (TX)
3rd Emma Cottini (NY)
3rd Kelly Schroeder (MT)

Source: www.princeton.edu/~jlogan/TKD/US-Nationals-Senior-files/1989-Sr.-National-results.htm

# 15th National Taekwondo Championships
Columbus, OH
April 13-15, 1989

## Black belt gyo-roogi

### Fin
**Men**
1st Juan Moreno (IL)
2nd Young Lee (OH)
3rd Hoang Ly (CA)
3rd Andrew Young (OH)

**Women**
1st Helen Yee (OH)
2nd Cheryl Kalanoc (CO)
3rd Cathy Gravelin (TX)
3rd Bonnie Watts (OH)

### Fly
**Men**
1st Luong Pham (OH)
2nd Hyon Lee (CO)
3rd Jeffrey Coffey (OK)
3rd John Monroe (TX)

**Women**
1st Mayumi Pejo (CO)
2nd Terry Poindexter (MO)
3rd Maria Gonzalez (NV)
3rd Thu Nguyen Smith (IN)

### Bantam
**Men**
1st Yun Won Jung (IL)
2nd Leon Lynn (WA)
3rd Donald Jackson (CO)
3rd Chris Berlow (NY)

**Women**
1st Diane Murray (CA)
2nd Ani Ahn (IL)
3rd Tammy Stamps (OK)
3rd Yolanda Bennett (SC)

### Feather
**Men**
1st Clay Barber (CO)
2nd Sung Kang (OK)
3rd Heath Watson (OH)
3rd Rodney Stum (IL)

**Women**
1st Kim Dotson (OH)
2nd Rose Chaplin (MA)
3rd Mary Horvith (CT)
3rd Ayoka Brown (DC)

### Light
**Men**
1st Garth Cooley (IN)
2nd Tim Connolly (CO)
3rd Richard Ahn (IL)
3rd Bobby Clayton (AY)

**Women**
1st Julie Werhnyak (IL)
2nd Stephanie Magid (UT)
3rd Angela Wolbert (OH)
3rd Kristin Ehrmantraut (ND)

### Welter
**Men**
1st Greg Baker (OH)
2nd Bobby Kim (NC)
3rd Brian Laney (OK)
3rd Jordan Perry (CA)

**Women**
1st Kristi Koch (IN)
2nd Teresa Baddour (OH)
3rd Anita Silsby (CO)
3rd Arlene Limas (IL)

### Middle
**Men**
1st Rory Vierra (CA)
2nd Herb Perez (NJ)
3rd Michael Pejo (CO)
3rd Victor Clark (NV)

**Women**
1st Lydia Zele (CA)
2nd Sharon Jewell (DC)
3rd Diana Mason (MD)
3rd Michelle Smith (NY)

### Heavy
**Men**
1st Scott Miranti (MT)
2nd Greg Tubbs (TX)
3rd George Weissfisch (TX)
3rd Emmett Tademy (MI)

**Women**
1st Kathy Wagner (CO)
2nd Cindy Anglemyer (MI)
3rd Rhonda Juarez (TX)
3rd Gwen Teague (TX)

Source: www.princeton.edu/~jlogan/TKD/US-Nationals-Senior-files/1989-Sr.-National-results.htm

# CHAPTER 10
# INTO THE '90S:
## THE EVOLUTION OF THE ARMED FORCES TAEKWONDO TEAM

The decade of the 1990s continued to demonstrate the growth of the Armed Forces Taekwondo Program. Change was definitely in the air at Fort Indiantown Gap, as part of the 1990 All-Army Taekwondo Trial Camp, a competition which many referred to as North American CISM or the North American Friendship Games.

The competition would be a match between members of the All-Army Taekwondo Team and the Canadian Armed Forces Taekwondo Team. There was an attempt to get the Mexican Armed Forces Team to join the competition, but unfortunately that was not possible, leaving only the US and Canada to square off against one another competing for the Team trophy.

According to the Army Taekwondo Association Newsletter, both teams had a total of ten competitors with two competitors per division. The weight classes were as follows: Lightweight, Welterweight, Middleweight and Heavyweight. The competition would be single elimination consisting of three two-minute rounds.

The Army's team consisted of four lightweights: Bobby Clayton, Reginald Perry, Jay Debouse and Gavin "Hutch" Hutchinson. Robert Laury and Lawrence Powell were the team's welterweights, Michael Baker and Luis Torres were the team's middleweights, and Scott Lekane and Dwayne "Dewey" Lopp rounded the team as the Army heavyweights.

There were 16 matches in total throughout the event, with the All-Army Team successfully defeating the Canadian Armed Forces Team with an overall score of nine wins for the Army and seven wins for the Canadian Armed Forces Team.

This competition was made a reality thanks to the efforts of George Garnett, Steve Brown, Paul Boltz, Claudia Berwager, and the Fort Indiantown Gap staff, which hosted the event. On the Canadian Armed Forces side of the house, it was Colonel Robert Martineau and his staff at the Canadian Combined Forces Military Command located in Ottawa, Canada.

The competition would continue until late 1997; why the competition was later cancelled remains unclear at the time of my research into this event. This competition at the time was a *big* step forward for the All-Army and the US Armed Forces Taekwondo programs overall.

It's definitely one helluva way to open the 1990s and this was only the beginning! Within this decade we also saw the birth of the All-Army and Armed Forces Women's Taekwondo Team.

The 1990 USTU National Championships would be held in the Midwest in the great state of Wisconsin. Madison, Wisconsin was the city chosen to host the event. The Army fielded a seven-soldier Taekwondo team. The men's team consisted of Donald Jackson in the Bantamweight Division, Gavin Hutchinson and Yong Kim in the Featherweight Division, Bobby Clayton and Reginald Perry in the Lightweight Division and lastly, Scott LeKane in the Heavyweight Division. Representing the women's team: Jennifer McGreggor in the Women's Welterweight division.

The team was reinforced by the soldiers who were not selected but were qualified to attend the event. Those soldiers were Darby Holsing, Jody Gibson, Barton Gonzales, Marie Garn and Jada Monroe.

This year's sparring championships took place over a two-day period with four of the eight weight classes competing each of the two days. Unfortunately, this small yet strong team did not perform as well as the previous two teams, becoming the All-Army and the Armed Forces programs' first real defeat since 1988.

Bobby Clayton withdrew from competition for undisclosed reasons, Jennifer McGreggor was also forced to withdraw due to a blood clot discovered in her foot, Donald Jackson was unable to progress after winning three matches, both Reginald Perry and Yong Kim lost their first match, Gavin Hutchinson was forced to withdraw due to a series of misdiagnosed hairline fractures found on his foot by the medical staff at Fort Indiantown Gap, and lastly, Scott LeKane was unable to compete at the tournament.

Despite these many setbacks, the Army team continued to press forward, accomplishing its goal of successfully placing a soldier on the US National Team. Jada Monroe became the All-Army Taekwondo Team's first soldier to earn a place on the US National Team, earning a bronze medal at the 1990 Pan-American Championships in the Women's Middleweight division.

Rafael Medina and Jay Debouse continued the spread of All-Army Taekwondo, and Pedro Laboy and Dwayne Lopp took center stage. Laboy earned a position as an alternate on the Puerto Rican national team. Rafael Medina began teaching Taekwondo while stationed in Germany and successfully registered his school under the USTU banner. Dwayne Lopp and Jay Debouse began to promote the team via the Armed Forces Network, also while stationed in Germany; these activities helped the program to gain further exposure.

Quite a few events had been planned for the month of August 1990; those selected for this year's Armed Forces Team were scheduled to compete in a series of friendship competitions with Korea, China, and Japan. The team was scheduled to compete at the Pan-American Championships in Puerto Rico, and Jada Monroe would make her debut as the first US female soldier to compete in the World Cup held in Spain.

Stateside, the Army Taekwondo Association had scheduled a competition at Fort Myer, Virginia to continue to promote the association and All-Army Taekwondo, in late October. Unfortunately, most of these scheduled events were canceled due to the invasion of Kuwait by the Iraqi military.

With Operation Desert Shield and Operation Desert Storm complete, the US armed forces slowly began to pick up where they left off. The February 1991 issue of the Army Taekwondo Association Newsletter offered insight into what was on the horizon for the team and the sport.

This year's calendar had several events scheduled. Those events were: the Elite Athletes Tournament hosted by the Olympic Training Center, The Army Taekwondo Association's Military Open to be held at Fort Myer, Virginia, the All-Army Taekwondo trial camp and fight-off at Fort Indiantown Gap, the North American Friendship tournament (aka North American CISM) hosted by Canada and the Canadian Armed Forces Taekwondo program, the USTU National Championships in Portland, Oregon, the US Team Trials hosted by the Olympic Training Center, the World Military Championships (CISM) in Seoul, Korea, The Olympic Sports Festival in Los Angeles, and the Pan-American Games in Havana, Cuba.

It was this year that an up-and-coming lightweight from Pickens, South Carolina named Reginald "Reggie" Perry became the 1991 All-Army Team's newest national champion, capturing the top slot at Portland, Oregon, with Jada Monroe capturing the silver in the Women's Middleweight division.

# 17th National Taekwondo Championships
Portland, OR
May 16-19, 1991

## Black belt gyo-roogi

### Fin

**Men**
1st Juan Moreno (IL)
2nd Shawn Durden (FL)
3rd Hoang Ly (CA)
3rd Ian Mitchell (CO)

**Women**
1st Bettina Bairley (FL)
2nd Diana Radakovic (CA)
3rd Sherry Getchman (NV)
3rd Kim Brown (FL)

### Fly

**Men**
1st Justin Poos (OK)
2nd Craig DeRosa (NY)
3rd Albert Levingston (CO)
3rd John Monroe (TX)

**Women**
1st Mayumi Pejo (CO)
2nd Selena Hernandez (NY)
3rd Kina Elyassi (MD)
3rd Gena Short (OH)

### Bantam

**Men**
1st Scott Fujii (CA)
2nd Leon Lynn (WA)
3rd Brian Yoon (IN)
3rd Young Lee (OH)

**Women**
1st Diane Murray (CA)
2nd Jennifer Grabel (FL)
3rd Kim Dotson (OH)
3rd Tammy Stamps (OK)

### Feather

**Men**
1st Jean Lopez (TX)
2nd Tuoi Nguyen (WA)
3rd Eui Lee (MN)
3rd Elbert Kim (CA)

**Women**
1st Darcy De Kriek (CA)
2nd Jennifer Srutowski (CA)
3rd Denise Piker (ND)
3rd Kelly Brady (OH)

### Light

**Men**
1st Reginald Perry (MA)
2nd Michael Gruetzmacher (WI)
3rd Nick Terstenjak (MT)
3rd Gordon White (VT)

**Women**
1st Delores Johnson (IN)
2nd Deb Farrell (IA)
3rd Kristi Koch (IN)
3rd Nancy Ferguson (CA)

### Welter

**Men**
1st James Villasana (TX)
2nd Daniel Miranda (FL)
3rd Michael Canada (OR)
3rd Gary Barron (CA)

**Women**
1st Arlene Limas (IL)
2nd Anita Silsby (CO)
3rd Antoinette Toto (NY)
3rd Kathy Yeh (CA)

### Middle

**Men**
1st Billy Petrone (CT)
2nd Rory Vierra (CA)
3rd Greg Baker (OH)
3rd David Weeks (CA)

**Women**
1st Sharon Jewell (DC)
2nd Jada Monroe (MD)
3rd Chavela Aaron (MA)
3rd Michelle Smith (NY)

### Heavy

**Men**
1st John Roche (FL)
2nd Scott Miranti (MT)
3rd Dimitri Diatchenko (FL)
3rd Emmett Tademy (MI)

**Women**
1st Lynette Love (DC)
2nd Kathy Wagner (CO)
3rd Christina Bayley (OH)
3rd Jenny Webb (IL)

The results of the 1991 USTU National Championships

As for the World Military Championships (CISM) in Seoul, Korea, the US Armed Forces did send participants to the competition. Who comprised this team, representing the US Armed Forces at this event, was unavailable at the time of my writing this book

## 3rd World Military Taekwondo Championships (CISM)
### Seoul, Korea
### June 15-21, 1991)

| Weight | Rank | Male | | Female | |
| --- | --- | --- | --- | --- | --- |
| | | Name | Nationality | Name | Nationality |
| Fin | 1 | Kang Chul Woo | KOR | | |
| | 2 | Michael | PHI | | |
| | 3 | Sabaruddin | MAL | | |
| | | Win Kyaw | MYA | | |
| Fly | 1 | Lim Chang Sup | KOR | | |
| | 2 | Bilia Massimo | ITA | | |
| | 3 | Feras Rafeo Jayuse | JOD | | |
| | | Aung Kyaw Sol | MYA | | |
| Bantam | 1 | Oh Yung Zoo | KOR | | |
| | 2 | Kouassi Guy Framck | CIV | | |
| | 3 | Jayausi Nasser | BEL | | |
| | | Snwe Tun | MYA | | |
| Feather | 1 | Kim Soo | KOR | | |
| | 2 | Isernia Nicola | ITA | | |
| | 3 | Poupolo Stephane | BEL | | |
| | | Yaser Al Ghazwi | YOR | | |
| Light | 1 | Yousuf Mohammed Abuzaed | YOR | | |
| | 2 | Abdulaziz K.Sweai | LBA | | |
| | 3 | Kim Byung Chul | KOR | | |
| | | Khamis Hamed Abbas | SUD | | |
| Welter | 1 | Remark Patrice Abdoulaye | CIV | | |
| | 2 | Zahediahashi Hassan | IRI | | |
| | 3 | Wedal a.s.Hureirat | JOR | | |
| | | Jin Jung Woo | KOR | | |
| Middle | 1 | Yoon Soon Chul | KOR | | |
| | 2 | Abrasi Seyed Hossein | IRI | | |
| | 3 | Ammar Fahed Sbeiz | JOR | | |
| | | Martin Kenneally | CAN | | |
| Heavy | 1 | No Sin | KOR | | |
| | 2 | Everald Wright | CAN | | |
| | 3 | Ehab F.M.Cabaritz | JOR | | |
| | | | | | |

# CHAPTER 11
# GAINING MOMENTUM

1992 All-Army Taekwondo Team

This year's All-Army Taekwondo trial camp would feature a friendship competition with the Canadian Armed Forces Taekwondo Team; the event would be hosted by Fort Indiantown Gap and the soldiers who would make up the 1992 All-Army Taekwondo Team prepared in earnest to defend their "house."

I was given a copy of video footage dating back to January 29, 1992. The footage begins with Bobby Clayton and Pedro Laboy training together at the Blue Mountain Gym at Fort Indiantown Gap. I am able to see firsthand just how formidable Bobby's abilities were back then.

Both he and Pedro seemed to be on equal footing in terms of technical knowledge, but Bobby's kicks seemed to appear smooth and effortless. Another section of the video shows these two men practicing target drills. Although the two of them seemed very relaxed, I can see just how technically advanced he was back then.

Pedro was practicing a technique called a skipping round kick, often referred to as a fast kick. From the fighting position, one would push off with the back leg in a skipping motion to deliver a kick with the lead leg. This is comparable to the jab in boxing, with the body being the primary target.

The video shifts to Bobby Clayton who executes a running kick, referred to as a "fast step" round kick. The technique is executed from the fighting position and is performed with a sudden burst of speed followed by a round kick to the body. Bobby looks relaxed and even in this relaxed state, he executes this technique almost effortlessly.

Now before anyone begins to make this assumption that I'm showing favoritism to Bobby, I'm going to stop you right there. I'm simply stating what I saw and how his fellow teammates viewed him at that time.

Make no mistake, the entire team was full of talented people to include the charismatic Reginald Perry, national champion and CISM medalist; Ms. Rachel Ridenour, who went on to become a CISM champion in the Women's Welterweight division and a US National champion in 1996. You had returning Army Heavyweight and pioneer of the Fort Bragg Taekwondo Team Pedro Laboy, then there was Mr. Jody Gibson, and future Fort Hood and 1998 All-Army Taekwondo Team coach, Mr. Michael Bennett.

Just seeing this video and these individuals during their years as competitors leaves me with a sense of pride and gratitude. It was these soldiers that paved the way for Soldier-athletes such as me. Any soldier past or present, that is competing in this sport representing the Army, in my opinion owes them the same because without them "stepping up and stepping out" (thank you, Mark Green) the program would have faded just as quickly as it began.

The video then shifts to the North American Friendship Exchange between the US and Canadian Armed Forces teams. The opening ceremonies were flawless demonstrations of pageantry, military discipline and Taekwondo excellence all rolled into a single event. Both teams were warmly greeted with a round of applause; however, since the Blue Mountain Sports Arena was the US Armed Forces' back yard, they were met with loud and thunderous applause.

US Lightweight champion Reginald "Reggie" Perry offered up a prayer of invocation to the two teams and to all who were involved. Lastly, the national anthems for both Canada and the US were played as a sign of mutual respect and continued friendship. Both teams stood at attention as each anthem played. Once the final bar of the US anthem was played, it was time to get down to business.

Coach Bruce Harris offered commentary regarding the competition, helping the spectators who weren't quite familiar with the rules of Taekwondo to gain a better understanding of the game. Much of his commentary was fused with comedic punchlines which helped to break some of the monotony that began to set in. A man stepped into the vert center of the competition ring and shouted "Chung! Hong!" (Korean for Blue and Red). The first of the day's competitors entered the competition area.

Beginning with the competitor wearing the blue chest-protector (Canada) and then red (US), he inspects them for proper protective gear. The two service member athletes then exchange gifts of friendship. The two competitors are ordered to attention and to turn facing the judges' table to bow in respect, followed by the two turning to face one another, rendering the same respectful bow.

Donald "Action" Jackson, true to form, lived up to his nickname. For three rounds, three minutes each round, there was plenty of action coming from Mr. Donald Jackson. Every kick thrown by Mr. Jackson landed with a loud, yet dull thud. However, at the end of the match Canada had claimed victory during this exciting and action-packed match. Up next was US National lightweight champion Mr. Reginald Perry. Reggie's style: that of a street brawler and he'd hit you with everything but the kitchen sink!

Mr. Perry's opponent learned what was to come during their match because it seemed that more than once, the Canadian competitor found himself being sent to the mat by Reggie's flurries of kicks and punches! During their second exchange Reggie, with pinpoint precision, sets his opponent up for an axe-kick trap by stepping back with his right leg, giving the impression that he was planning to execute a right leg roundhouse kick. The Canadian fighter responds with a right leg round kick of his own, only to be beaten with a right leg axe-kick from Reggie.

In the background there were continued shouts of support from each team, "USA!" "Canada!" "USA!"

"Canada!" "USA!" "Canada!" Even a simple team chant became a fierce yet friendly rivalry. During another of their exchanges where Canada was sent to the mat, Reggie rendered a hand gesture signaling "get the f**k outta here with that mess!"

Reggie, fueled by that knockdown, kept the momentum moving forward into the remaining seconds of the final round, with his kicks becoming more and more accurate, ensuring that he would emerge victorious at the end of the match. Finally, it was time for the main attraction: Lightweight, US National Champion and Team Captain of the Armed Forces team, Mr. Bobby Clayton, vs Canadian Lightweight National Champion Higgins.

The camera focuses on Bobby, who's wearing the red chest-protector, as he calmly moves to lightly warm up and turns to face his coach, rendering a sharp salute, executes an about face and moves towards the center of the mat. After he bows to both the judges' table and to his opponent, Mr. Higgins, both men assume their fighting positions.

The referee signals the match to begin and with a piercing shout Bobby begins to demonstrate his surgical precision. Four perfectly executed kicks from Bobby go unanswered by Higgins:

the first two were round kicks to the body, the other to the head followed by an axe kick to the head. Throughout this match Bobby Clayton put on a Taekwondo clinic with his opponent, Mr. Higgins, being reduced to target practice.

No matter what attack Higgins executed, Bobby had a perfectly timed answer. In fact, Higgins was nearly knocked out by a spin-hook kick executed by Bobby. It was clearly obvious that Bobby was the superior athlete during this match. At the end of the match Bobby emerged victorious. It was the All-Army team's finest hour!

During the 1992 season, the Armed Forces claimed two medals. In the Women's Lightweight division, representing the US Navy, Elizabeth "Liz" Evans captured the top slot. In the Men's Heavyweight division Lieutenant Sean Burke of the US Marine Corps captured the first gold medal, becoming the Marines Corps' first national medalist.

The 4th World Military Taekwondo Championships (CISM) took place on October 17, 1992, in Tehran, Iran. However, due to the current relations between the two nations, the US Armed Forces did not send a team to the event.

The team came up empty at the 1993 US National Championships in St. Paul, Minnesota, but the 5th World Military Championships became the US Armed Forces Team's redemption, with Reginald Perry capturing a bronze medal in the Men's Lightweight division.

### 18th National Taekwondo Championships
Hampton Coliseum
Hampton, VA
May 28-30, 1992

**Fin**

| Men | | | Women | |
|---|---|---|---|---|
| Shawn Durden | FL | 1st | Shannon McCuin | CA |
| Hoang Ly | CA | 2nd | Michelle Ab | CA |
| Hyon Lee | CA | 3rd | Vicki Slane | OH |
| Ian Mitchell | CO | 3rd | Shannon Roslan | IL |

**Fly**

| Men | | | Women | |
|---|---|---|---|---|
| Angel Aranzamendi | MI | 1st | Terry Poindexter | MO |
| Dennis Pinaroc | TX | 2nd | Yasmin Smith | WA |
| Justin Poos | OK | 3rd | Sayun Kelly | CO |
| Lai Vo | NY | 3rd | Ari Ahn | IL |

**Bantam**

| Men | | | Women | |
|---|---|---|---|---|
| Joey Febres | CA | 1st | Janet Yi | VA |
| Erick Kenley | CA | 2nd | Darcy DeKriek | CA |
| Sammy Pejo | CO | 3rd | Michelle Thompson | NC |
| Clayton Barber | TX | 3rd | Kim Dotson | CO |

**Feather**

| Men | | | Women | |
|---|---|---|---|---|
| Mark Williams | NJ | 1st | Kim Wakefield | MN |
| Dong S. Lee | TX | 2nd | Simona Hradil | TX |
| Gary McFeeders | OH | 3rd | Jennifer Srutowski | IL |
| Keith M. Jones | CA | 3rd | Tiffany Norris | IL |

**Light**

| Men | | | Women | |
|---|---|---|---|---|
| Kevin Padilla | NJ | 1st | Nancy Ferguson | CA |
| Garth Cooley | IN | 2nd | Elizabeth Evans | WA |
| Nick Terstenjak | MT | 3rd | Eleni Bardatsos | NY |
| James Villasana | TX | 3rd | Jennifer Fischer | NY |

**Welter**

| Men | | | Women | |
|---|---|---|---|---|
| Ken Hance | NC | 1st | Chavela Aaron | MA |
| Jimmy Graesser | CO | 2nd | Danielle Laney | OK |
| Bobby Smith | HI | 3rd | Sharon Jewell | CO |
| Bobby Kim | NC | 3rd | Heather Tallman | MT |

**Middle**

| Men | | | Women | |
|---|---|---|---|---|
| Hector Barrios | FL | 1st | Charity Girt | FL |
| Billy Petrone | CT | 2nd | Renee Martinez | PA |
| Nick Charavelle | CA | 3rd | Laurie J. Blum | NY |
| Paul Reyes | NV | 3rd | Monica Macisak | WI |

**Heavy**

| Men | | | Women | |
|---|---|---|---|---|
| Sean Burke | NJ | 1st | Mary Friswell | DE |
| Paris Amani | CO | 2nd | Robin Umphrey | CA |
| Michael Pejo II | CO | 3rd | Christina Bayley | OH |
| Owen Brown | CT | 3rd | Jenny Webb | IL |

As the 1994 season began, the Army team and the Armed Forces team, respectively, received some new blood to bolster their team rosters: 1991 Lightweight National Champion Reginald Perry returned to the team followed by Army Heavyweight Edward Givans, newcomers Bongseok Kim fighting in the men's Bantamweight division, Schileen "Weenie" Potter fighting in the women's Bantamweight division, Rachel Ridenour representing the women's

Welterweight division, and Pedro Cruz-Febo fighting in the men's Welterweight division.

For the Armed Forces, competing in the women's Lightweight division, returning National Champion Elizabeth Evans represented the US Navy. In the Men's Featherweight division, Ron Onyon represented the US Air Force.

Sadly, this would be Bruce Harris' final year as head coach of the Army and Armed Forces Taekwondo Team. Bruce stepped down as head coach to maintain and progress his military career, and former Army Taekwondo Team Captain Bobby Clayton was selected to succeed Bruce as head coach.

Initially, Bobby was recruited by WCAP Sports Director Mr. Havlicek. I later learned that at the request of former All-Army Taekwondo Heavyweight Edward Givans, Bobby became the first full-time head coach of the Taekwondo WCAP program. Bobby's experience as an athlete and his knowledge of the modern Taekwondo game as it was played by the Koreans made him the perfect choice as head coach for the program.

Shortly after accepting the position, he was later recruited by Mr. Boltz to become the head coach of the Army Taekwondo team as well. Armed with the knowledge he'd received while training at Dongsan High School, Bobby began to implement what would become the Army team's new training regimen, which included the Dongsan High School's style of Taekwondo.

In addition to his new direction, he selected former All-Army Bantamweight and CISM medalist Rafael Medina as his assistant coach. Bobby had a very clear vision of how he wanted both programs to operate. In his vision the All-Army Taekwondo Team would serve as a military grassroots program where the athletes would train with him for the duration of the trial camp. Based upon their progression, they would potentially have the opportunity to train with him for up to three months.

Those who were selected for WCAP would train with him full time throughout a three-year timeframe.

During this time, Bobby, while developing the next generation of athletes, also developed the talent, knowledge, and skill of his assistant coach Rafael Medina, who would later go on to become the 1998 and 1999 Armed Forces Taekwondo Team head coach.

Bobby utilized the WCAP guys to integrate the program's system of knowledge and used them to the advantage of All-Army athletes by mixing the two groups. After a short time, the All-Army team (in terms of knowledge and skill) caught up with WCAP, mainly due to one person who played a critical role in closing that knowledge gap.

Sergeant Michael R. Bennett, who was competing as a heavyweight at that time, took the new All-Army Taekwondo training regimen created by Bobby back to Fort Hood, Texas and took over as head coach of the Fort Hood Taekwondo Team. He began training soldiers who comprised the Fort Hood Taekwondo Team in this new All-Army regimen.

The Fort Hood program mimicked the environment used by WCAP and the Fort Hood soldiers were allowed to train full time under Coach Bennett. In 1996, Bennett brought Howard Clayton, Kevin Williams, and Andrew Roberts to the All-Army Trial Camp, which resulted in both Andrew and Kevin being selected as members of the 1996 team.

With the 7th World Military Championships being held in Isfahan, Iran, the US did not send a team to compete. However, Armed Forces team players Elizabeth Evans, Rachel Ridenour, and Sean Burke captured three gold medals at the USTU National Championships held in Colorado Springs.

## 5th World Military Taekwondo Championships (CISM)

Date: June 6-15, 1993
Place: Royal Military College, Kingston, Ontario, Canada

| Participating Nations | 12 (Belgium, Canada, Germany, Greece, Iran, Italy, Korea, Kuwait, Lebanon, Peru, Saudi Arabia, USA) |
|---|---|

| Weight | Rank | Male | | Female | |
|---|---|---|---|---|---|
| | | Name | Nationality | Name | Nationality |
| Fin | 1 | Jeong Woon Lee | KOR | | |
| | 2 | Abdulla Al Assiri | KSA | | |
| | 3 | Eamon Nolan | CAN | | |
| | | Loghman Kesheshavarz | IRI | | |
| Fly | 1 | Hwa Jin Lee | KOR | | |
| | 2 | Andreas Krog | GER | | |
| | 3 | Galal Chorooki | IRI | | |
| | | Mohammed Sadiq | KSA | | |
| Bantam | 1 | Asghar Tahmasebi | IRI | | |
| | 2 | Ariel Del Rosario | CAN | | |
| | 3 | Panagiotis Magalios | GRE | | |
| | | Ki Jeong Park | KOR | | |
| Feather | 1 | Jong Hyun Lee | KOR | | |
| | 2 | Stepgen Bilodeau | CAN | | |
| | 3 | Mohsen Raffi | IRI | | |
| | | Rudy Bucovaz | BEL | | |
| Light | 1 | Ercan Ozkuru | GER | | |
| | 2 | Eui Chun Sung | KOR | | |
| | 3 | **Reginald Perry** | **USA** | | |
| | | Khaled Al Shamrani | KSA | | |
| Welter | 1 | Dong Hee Lee | KOR | | |
| | 2 | Woled Al Blyshi | KSA | | |
| | 3 | Luis Zambrano | PER | | |
| | | Ebvahim Saadati | IRI | | |
| Middle | 1 | Seyed Hossein Abbasi | IRI | | |
| | 2 | Zu Suk Jung | KOR | | |
| | 3 | Andrew Ross | CAN | | |
| | | Giacomo Grappiolo | ITA | | |
| Heavy | 1 | Yoon Ki Hong | KOR | | |
| | 2 | Massimiliano Romano | ITA | | |
| | 3 | Hassan Aslani | IRI | | |
| | | Olaf Wilkens | GER | | |

# 6th World Military Taekwondo Championships (CISM)

Date: September 3-14, 1994
Place: Peruvian Military Academy, Lima, Peru

Participating Nations　　Men - 8 (Belgium, Canada, Korea, Kuwait, Peru, Russia, Saudi Arabia, USA)
　　　　　　　　　　　　Women - 4 (Canada, Korea, Peru, USA)

| Weight | Rank | Male | | Female | |
|---|---|---|---|---|---|
| | | Name | Nationality | Name | Nationality |
| Fin | 1 | Abdullash Al-Assri | KSA | | |
| | 2 | Rodriguez Alvarado Angel | PER | | |
| | 3 | Young Kil JUNG | KOR | | |
| | | Tran Minh Juan | CAN | | |
| Fly | 1 | Jae Ki JI | KOR | Liza Lopez | Canada |
| | 2 | Tanaka Yamasato Joao Manul | PER | | |
| | 3 | Mohammed A. Saddiq | KSA | | |
| | | Choon Joul YANG | KOR | | |
| Bantam | 1 | Sang Jun JUNG | KOR | **Potter Schileen** | **USA** |
| | 2 | Lamas Olmos Cesar | PER | | |
| | 3 | Jean Lavoie | CAN | | |
| | | **Kim Bongseok** | **USA** | | |
| Feather | 1 | Dae Jin JANG | KOR | Rivasplata Cavila | Peru |
| | 2 | Comert Mehmet | BEL | | |
| | 3 | Morales Berrocal Sergio | PER | | |
| | | **Pedro Oyon Ronald** | **USA** | | |
| Light | 1 | **Perry Reginald** | **USA** | **Elisabeth Evans** | **USA** |
| | 2 | Bucovaz Rudy | BEL | | |
| | 3 | Adam Tremblay | CAN | | |
| | | Choon Joul YANG | KOR | | |
| Welter | 1 | Jae Chun CHO | KOR | **Ridenour Rachel** | **USA** |
| | 2 | Cavalie Chavez Abel Antonio | PER | | |
| | 3 | Nyron Higgins | CAN | | |
| | | **Cruz-Febo Pedro** | **USA** | | |
| Middle | 1 | Joo Suk JUNG | KOR | | |
| | 2 | Faisal M. Hakami | KSA | | |
| | 3 | Melgar Sheen | PER | | |
| | | Roberto Andy Ross | CAN | | |
| Heavy | 1 | Mravo Mejia Gonzalo | PER | | |
| | 2 | Jong Bum PARK | KOR | | |
| | 3 | MArtin Kenneally | CAN | | |
| | | A. Karpov Alexei | RUS | | |

# 22nd National Taekwondo Championships
## Olympic Training Center
## Colorado Springs, CO
## May 10-12, 1996

## Black belt gyo-roogi

### Fin

**Men**
1st David Montalvo (TX)
2nd Gilbert Johnson (OH)
3rd Lyle Baldonado (HI)
3rd Jaime Mystic (AF)

**Women**
1st Kay L. Poe (TX)
2nd Taryn Roman (NY)
3rd Cynthia Morgan (IN)
3rd Kenia Sosa (NY)

### Fly

**Men**
1st Ruben Gayon (WI)
2nd Justin Poos (OK)
3rd David Bartlett (NY)
3rd James Stagen (IL)

**Women**
1st Mandy Meloon (LA)
2nd Kimberly Matsunaga (CO)
3rd Julie Harris (IL)
3rd La-Tina Snyder (IL)

### Bantam

**Men**
1st Antony Graf (NY)
2nd Craig DeRosa (NY)
3rd Steven Lee (NJ)
3rd Boris Zeissig (CA)

**Women**
1st Sanaz Shahbazi (CA)
2nd Tracy Wiliamson (NC)
3rd Kathy Araneta (CA)
3rd Yolanda Bennett (SC)

### Feather

**Men**
1st Steven Lopez (TX)
2nd Raphael Park (WA)
3rd Wayne DeRosa (NY)
3rd Jin W. Suh (NY)

**Women**
1st Debra Palazzi (NC)
2nd Jennifer Mohammed (NY)
3rd Kathryn Cosentino (NY)
3rd Kimberly Gayon (WI)

### Light

**Men**
1st William Henson (TX)
2nd Glenn Lainfiesta (CA)
3rd Joe Ash (VA)
3rd Kevin McCullough (CA)

**Women**
1st Elizabeth A. Evans (WA)
2nd Delores Johnson (IN)
3rd Simona Hradil (TX)
3rd Kelly Thorpe (OK)

### Welter

**Men**
1st Troy Garr (CA)
2nd Javier Sanchez (TX)
3rd Clyde Gordon (CA)
3rd Vincent Nguyen (TX)

**Women**
1st Rachel Ridenour (AY)
2nd Holly Fuezy (CA)
3rd Jennifer Jose (MA)
3rd Kellie Morse (MA)

### Middle

**Men**
1st Peter Bardatsos (NY)
2nd Gregory Tubbs (TX)
3rd Paul Nelson (AY)
3rd Chris Zapata (CA)

**Women**
1st Barbara Kunkel (CO)
2nd Collette Chambers (MD)
3rd Jovita Brantley (WI)
3rd Rachael McQuarrie (CT)

### Heavy

**Men**
1st Sean Burke (NJ)
2nd Ryon Frederick (FL)
3rd Charles Alexander (CO)
3rd Michael Saincrez (IL)

**Women**
1st Miranda Hinrichs (IA)
2nd Yukie Ozawa (FL)
3rd Patricia Lewis (IA)
3rd Sharon Williams (FL)

# CHAPTER 12
# MICHAEL BENNETT AND THE FORT HOOD TAEKWONDO TEAM

1996 Fort Hood Taekwondo Team

I could not in good conscience write about the history of the Armed Forces team without spending some time writing about Mr. Bennett and his abilities as a forward-thinking coach. To date, he has been responsible for the largest group of potential athletes coming from Fort Hood, Texas, and he helped pioneer what would become a reboot of the 2nd Infantry Division's Taekwondo Team.

Michael R. Bennett, "Coach" to those who were trained by him, was competing as a Heavyweight with the All-Army Taekwondo Team. Bennett's first appearance at the All-Army trial camp initially began in 1991. While stationed at Fort Bliss, Texas, he came under the guidance of former All-Army Bantamweight and future All-Army Taekwondo Team Assistant Coach Rafael Medina. Michael's drive and determination to become one of the best led him to join Rafael Medina and future Army Team Head Coach Bobby Clayton in South Korea to continue his training in 1993.

In 1995, Bennett was asked by Coach Clayton to go to Fort Hood and take charge of the Fort Hood Taekwondo Team. Armed with the knowledge he'd received from Coach Clayton, he assumed the position as the head coach of the post team. Bennett also demonstrated a unique ability to network with the right people in key leadership positions within the 2nd Armored Division and Fort Hood's main division III Corps.

Prior to Bennett's arrival at Fort Hood, the post team was already established and at the time the team had current members of the All-Army Taekwondo Team. Comprising the team: Howard Clayton, Mark Lucas, Andrew Roberts, John Wetzel, and up-and-coming All-Army hopeful Kevin Williams. Mark was preparing to transfer to another post, and he briefed Bennett on his efforts to grow the team and to train the next generation.

The support that he was able to gain from both III Corps and the 2nd Armored Division's leadership resulted in Bennett being able to restructure the Fort Hood Taekwondo Team to mimic the Army World Class Athlete Program (WCAP), where full-time members of the team would be released from their units to train full time with Bennett in preparation for the All-Army Taekwondo trial camp.

Bennett's method for attracting new talent came in the form of an article in the Fort Hood newspaper, "Fighters wanted, will train; no experience necessary"; this, along with attending local open-style martial arts tournaments and visiting a few of the local martial arts schools frequented by soldiers stationed at Fort Hood.

Bennett began to restructure how the team trained using the morning and afternoon practice as a means to continue to improve the skill sets of Clayton, Roberts, Wetzel, and Williams, and the evening practice for newcomers to the team. One of these evening practices attracted the attention of John Swan, who began to train with the team as a welterweight.

By early 1996, Bennett along with Clayton, Roberts, Wetzel, and Williams, departed for the 1996 All-Army Taekwondo trial camp. At the end of the trial camp, Clayton, Roberts, and Bennett were initially selected as members of the 1996 Army Team; however, Bennett gave his position as team Heavyweight to Williams and would join the Army at the National Championships in Colorado Springs later that year. Bennett's primary focus was to return to Fort Hood and continue to prepare the next generation of All-Army athletes and prepare for the nationals on his own. Bennett was later joined by Army Middleweight Todd Angel, whom Bennett appointed as his assistant coach. This new group of potential athletes consisted of John Swan, Nicolau Andradae, Louis Davis (me), Ryan Lundy, London Arevalo, Michael Booker, Joseph Cruz, Luis Baretto, Tim Odie, and Petra Kaui. With this emerging nucleus of talented athletes, Bennett began the process of training and refining the Fort Hood Team.

That refinement was a grueling series of strength and conditioning drills taught to him by Coach Clayton.

The morning practice consisted of a 3.5-mile run designed to simulate the conditions that the athletes would face at Fort Indiantown Gap. At the end of the run, the athletes would perform a series of wind sprints to further increase lung capacity and build the explosive speed and power necessary to become a Taekwondo athlete.

Lastly, the athletes would be subjected to a grueling ab routine which required them to perform 500 crunches before training would end. These weren't merely training tools; they were also designed to sharpen their mental focus during the fatigue that they would experience during competition, forcing them to think clearly and choose their moves carefully.

The afternoon practice was headed by Assistant Coach Todd Angel, who placed even further emphasis on strength training and conditioning, target acquisition and accuracy, and situation drills *(Hogu* Drills). This helped the athletes build a repertoire, conventional and personal techniques, and the remainder of the afternoon practice consisted of sparring.

Sometimes Bennett would film these sessions with his video camera and the team would review the footage as a means to both critique and improve the growing skill sets of the athletes. Bennett's coaching style was one that never offered any compliments to the athletes, which kept the athletes focused and hungry, yet humble.

As time passed, the team got its first taste of competition when Fort Hood hosted its first Taekwondo championships. This event allowed the athletes to see how far they, as a team, had come and what needed more work. At the end of the competition, the entire team had medaled (placed within the top four) at the event. Although the athletes felt proud of their accomplishments, that pride was promptly dissected by Bennett during the video review of their performance. By mid-1996, the performance of the Fort Hood team caused a slight rivalry between the team's post Sports Director Mr. Ron Foster, and former post Taekwondo instructor James McMurray.

Unfazed by the friendly rivalry between these two, the team continued to move forward in their efforts to prepare for the 1997 All-Army season. In a continued effort to spread awareness of the All-Army Taekwondo program, Bennett brought the team to his hometown of Shreveport, Louisiana. At the request of friend and high school ROTC instructor, Army Major Ivory Irvin, the team staged a Taekwondo demonstration for the students of Fair Park High School, where Major Irvin was an instructor.

At the end of the demonstration, each of the Fort Hood team members received a letter of commendation from Major Ivory to be presented to their chain of command. During the team's visit to Bennett's hometown, they were hosted by Bennett's mother, who exposed them to the true meaning of Southern hospitality.

It was during this experience that the team bonded as a family, a bond that continues to this very day. Upon returning to Fort Hood, the team began to receive selection notices for the up-and-coming 1997 All-Army Trial Camp and nearly the entire team was included, save

Baretto and Booker, whose training wasn't consistent enough to remain on the team, and London, Petra, and Tim (who were civilians at that time). The team that accompanied Bennett to The Gap were Assistant Coach Todd Angel, Howard Clayton, John Swan, Nicolau Andradae, myself (Louis Davis), and Ryan Lundy.

As the date to depart for the trial camp drew closer, up ramped the intensity of the team's training which also included weight checks conducted twice daily. Bennett continued to scrutinize even the slightest of mistakes made by the potential athletes, constantly reminding them of the possibilities of what could happen were they to continue making these mistakes.

Finally, the day came when the team left for Fort Indiantown Gap, Pennsylvania. From the first training session, the Fort Hood team quickly demonstrated their knowledge of the modern Taekwondo game.

During the seminar conducted by Coach Clayton, members of the Fort Hood Team were told to remain silent when Coach Clayton asked other athletes attending the trial camp a series of questions, questions that the Fort Hood team knew thanks to Michael Bennett.

# CHAPTER 13
# THE PARADIGM SHIFT OF THE 1997 ARMED FORCES TEAM

1997 All-Army Taekwondo Team

The 1997 season would be run slightly differently this year due to a limited budget; this year's camp would also serve as both the All-Army Taekwondo Team Trials and the Armed Forces Championships. Coach Bobby Clayton and Rafael Medina were officially in charge of the camp overall. Coach Steve Harrington brought the Air Force Taekwondo Team, small in number yet a strong group of candidates, to include the likes of Kevin Jones, Ron Onyon and Reynaldo Martinez, USMC headed by Luis De La Rosa, and brought back team heavy-hitter Sean Burke.

Upon their arrival, those vying for positions on the 1997 All-Army Taekwondo Team were welcomed by the support staff of Fort Indiantown Gap: Coach Clayton, Coach Medina, and Mr. Boltz. This included their apparel for the trial camp which consisted of several pairs of shorts, several T-shirts and one hooded sweatsuit. Since this was a condensed trial camp, there were four practices each day with the first one beginning at 5:30 a.m.

The members of the Fort Hood Taekwondo Team, their abilities and how well they were prepared for the camp, seemed to impress quite a few people. Their hard work and dedication paid off at the Armed Forces Championships later that weekend.

Finally, the stage was set for what would be the most highly contested Armed Forces Championships since the very first camp held at Little Creek, Virginia a decade earlier. Once again, Blue Mountain Sports Arena became the battleground that decided who would represent the US Armed Forces at the World Military Championships in Italy later that year.

Bruce Harris returned to The Gap as officiator for the event, bringing along with him sanctioned USTU referees, ensuring that the current USTU and World Taekwondo Federation rules would be enforced during the competition. Reginald Perry returned as a guest scorekeeper. As the matches began, it seemed like nearly every weight division had some tough competition.

At the end of the competition a strong US Armed Forces and All-Army Taekwondo Team had been selected. As a testament to Michael Bennett's leadership and organizational skills, Todd Angel's knowledge of strength and conditioning, the following members of the Fort Hood team to include Bennett himself were selected: Nicolau Andradae, Louis Davis, and John Swan (selected as an alternate), along with Fort Hood Taekwondo alumni Andrew Roberts and Kevin Williams.

With little time to prepare for the 23rd USTU National Championships, the Army Taekwondo Team increased their intensity, constantly tweaking and refining the skill set of its newly selected team. Overall, this year's camp, even with a short timeframe to select, train and field a strong team representing the US Armed Forces, spoke volumes about the US military's ability to do more with less.

The time came to depart for Oakland, California, the site of this year's national championships. The athletes and coaching staff were bristling with a mix of humility, confidence and excitement as the team made its way from Pennsylvania to California. The first phase of the competition was to ensure that everyone competing at this event maintained their weight requirements for their respective divisions. It's just as nerve-wracking as waiting for one's respective division to be called during competition.

Try to imagine standing in a *very* long line with dozens of other athletes, next try to imagine the sounds of rumbling stomachs because each of you had to put off eating just to ensure that your weight was within the weight limit. The discipline that is drummed into every US armed services member during boot camp, helped them to easily make weight.

The local fast-food restaurants were filled nearly to capacity once the athletes made their weight, so the reward was surprisingly a pig-out session at one's favorite fast-food joint (usually McDonalds or Burger King). At the opening ceremony, the US Armed Forces Team showed its colors, demonstrating its commitment to excellence by executing a flawless display of drill and ceremony maneuvers; it truly was a sight to see.

Then it was time to get down to business. The Army captured its first medal of the event with Bongseok Kim capturing a bronze medal in the Men's Bantamweight division. Next the Navy's Elizabeth Evans would capture the top slot in the Women's Lightweight division, followed by the Army's Alisha Williams securing the Army's second medal, winning the gold in the Women's Welterweight division.

In the Men's Middleweight division, Paul Nelson captured the Army's third medal securing a bronze, and in the Men's Heavyweight division Michael Bennett, under the guidance of Coach Rafael Medina, secured a fourth medal, winning the bronze due to having to withdraw after sustaining a pulled hamstring during the semifinals.

Rafael Medina poses with the Army's newest national medalists.

Michael Bennett (far right) takes his place on the medalists' podium.

# 23rd National Taekwondo Championships

Oakland Convention Center
Oakland, CA
May 7-10, 1997

## Gyo-roogi

### Fin

| Men | | | Women | | | Men | | | Women | |
|---|---|---|---|---|---|---|---|---|---|---|
| David Montalvo | TX | 1st | Brittany Goellner | TX | | George Weissfisch | TX | 1st | Eryn Redmon | WA |
| Mario Reyes | IL | 2nd | Taryn Roman | NY | | David Wilbur | MA | 2nd | Dora McCarty | OH |
| Christopher Schmidt | CO | 3rd | Kay Poe | TX | | Charles Alexander | CO | 3rd | Faith Dougherty | WA |
| Quintin Eng | CA | 3rd | Marlyn Ortiz | FL | | Michael Bennett | AY | 3rd | Daisy Murdoc | TX |

(Heavy category — right side)

### Fly

| Men | | | Women | |
|---|---|---|---|---|
| Jason Torres | TX | 1st | Mandy Meloon | LA |
| David Bartlett | NY | 2nd | Alexandra Jacobs | UT |
| Michael Tan | CA | 3rd | Melody Rosada | FL |
| James Stagen | IL | 3rd | Shan-Yuan Ho | MA |

### Bantam

| Men | | | Women | |
|---|---|---|---|---|
| Peter H. Kim | OR | 1st | Diana Ciocan | CA |
| Sang Rowand | WA | 2nd | Kristina Brooks | MD |
| Craig DeRosa | NY | 3rd | Heather Larsen | MI |
| Bong Seok Kim | AY | 3rd | Michelle Thompson | NC |

### Feather

| Men | | | Women | |
|---|---|---|---|---|
| John Eing | CA | 1st | Jennifer Srutowski | CA |
| Steve Rosbarsky | MT | 2nd | Debra Palazzi | NC |
| Daren E. Lee | MI | 3rd | Yolanda Bennett | SC |
| Tony Smith | TX | 3rd | Jenny Malecki | CA |

### Light

| Men | | | Women | |
|---|---|---|---|---|
| Nick Terstenjak | MT | 1st | Elizabeth A. Evans | WA |
| Kevin McCullough | CA | 2nd | Kelly Thorpe | OK |
| David Kang | CA | 3rd | Ani Ahn | IL |
| James Park | CA | 3rd | Simona Hradil | TX |

### Welter

| Men | | | Women | |
|---|---|---|---|---|
| Nico Davis | CO | 1st | Aisha Williams | RI |
| Andre Victorian | DC | 2nd | Lory Dance | VA |
| Alex Casanova | CA | 3rd | Holly Groman | FL |
| Tim O'Connell | MA | 3rd | Amy Lee | TX |

### Middle

| Men | | | Women | |
|---|---|---|---|---|
| Peter Bardatsos | NY | 1st | Erica Hodge | NV |
| Sherman Spinks | DC | 2nd | Paige LaRose | WA |
| Stewart Gill | OH | 3rd | Gayle Larsen | |
| Paul Nelson | AY | 3rd | Sabrina Desormes | NJ |

With the 1997 Nationals in the history books, the Army and Armed Forces team focused its efforts to prepare its medalists for the upcoming US Team trials held in Phoenix, Arizona. Even with this strong group of national medalists and champions, the Army and Armed Forces team faced stiff competition from the best athletes in the country. The following are the official results of that hotly contested event.

## 1997 US National Team Trials
Phoenix, AZ
June 13, 1997

### Fin

**Male**
- 1st Yung S. Han (IL)
- 2nd Gilbert Johnson (OH)
- 3rd David Montalvo (TX)
- 3rd Mario Reyes (IL)

**Female**
- 1st Kay Poe (TX)
- 2nd Brittany Goellner (TX)
- 3rd Taryn Roman (NY)
- 3rd Linda Radakovic (CA)

### Fly

**Male**
- 1st Jason Torres (TX)
- 2nd Angel Aranzamendi (MI)
- 3rd James Stagen (IL)
- 3rd Michael Tan (CA)

**Female**
- 1st Mandy Meloon (HI)
- 2nd Elizabeth King
- 3rd Shan-Yuan Ho (MA)

### Bantam

**Male**
- 1st Steven Lee (NJ)
- 2nd Antony Graf (NY)
- 3rd Sang Rowand (WA)
- 3rd Bong Seok Kim (AY)

**Female**
- 1st Tracy Williamson (NC)
- 2nd Diana Ciocan (CA)
- 2nd Sanaz Shahbazi (CA)
- 3rd Michelle Thompson (NC)

### Feather

**Male**
- 1st Steven Lopez (TX)
- 2nd John Eing (CA)
- 3rd Wayne DeRosa (NY)
- 3rd Daren Lee (MI)

**Female**
- 1st Debra Palazzi (NC)
- 2nd Yolanda Bennett (SC)
- 3rd Jennifer Srutowski (CA)
- 3rd Jenny Malecki (CA)

### Light

**Male**
- 1st David Kang (CA)
- 2nd Joe Ash (VA)
- 3rd Kevin McCullough (CA)
- 3rd William Henson (TX)

**Female**
- 1st Simona Hradil (TX)
- 2nd Elizabeth Evans (WA)
- 3rd Kelly Thorpe (OK)

### Welter

**Male**
- 1st Troy Garr (CA)
- 2nd Jean Lopez (TX)
- 3rd Timmy O'Connell (MA)
- 3rd Andre Victorian (DC)

**Female**
- 1st Dana Martin (CA)
- 2nd Alisha Williams (CO)
- 3rd Holly Gromann (FL)
- 3rd Gigi DeMita (OH)

### Middle

**Male**
- 1st Peter Bardatsos (NY)
- 2nd Sherman Spinks (DC)
- 3rd Paul Nelson (AY)
- 3rd Chris Zapata (CA)

**Female**
- 1st Barbara S. Kunkel (WA)
- 2nd Erica Hodge (NV)
- 3rd Sabrina Desormes (WI)
- 3rd Collette Chambers (MD)
- 3rd Gayle Larsen

### Heavy

**Male**
- 1st David Wilbur (MA)
- 2nd Charles Vernon
- 3rd George Weissfisch (TX)
- 3rd Ryon Frederick (FL)

**Female**
- 1st Christina Bayley (OH)
- 2nd Dora McCarty (OH)
- 3rd Eryn Redmon (WA)
- 4th Faith Dougherty (WA)

One final mission remained before closing out the 1997 competition season: the 9th World Military Taekwondo Championships (CISM), which would take place in Ariccia, Italy in early October. The stakes for this event were high due to the number of countries fielding their best military athletes. The countries in attendance of this event aside from the US were Belgium, Canada, Ivory Coast, Croatia, Cyprus, Gabon, Germany, Greece, Iran, Italy, South Korea, Latvia, Lesotho, the Netherlands, Russia, Saudi Arabia, and Syria. A total of 18 countries competed for the title of military world champion.

During this event the US Armed Forces women's team took center stage, capturing medals in four of the eight weight classes in the sport. Elizabeth Evans continued her domination of the Lightweight division, winning the gold medal for the Navy, with Alisha Williams, C. Bales and Alana Conley winning bronze medals for the Army in the Women's Welter, Middle and Heavyweight divisions.

### 9th World Military Taekwondo Championships (CISM)
Date: October 5-10, 1997
Place: Nettuno, Ariccia, Italy

Participating Nations: 18 (Belgium, Canada, Cote d'Iviore, Croatia, Cyprus, Gabon, Germany, Greece, Iran, Italy, Korea, Latvia, Lesotho, Netherlands, Russia, Saudi Arabia, Syria, USA)

| Weight | Rank | Male | | Female | |
|---|---|---|---|---|---|
| | | Name | Nationality | Name | Nationality |
| Fin | 1 | K. Hong Il | KOR | V. Zoobkova | RUS |
| | 2 | V. Negichkin | RUS | M. Lemphane | LES |
| | 3 | G. Al Schumrani | KSA | | |
| | | Mehri M. | IRI | | |
| Fly | 1 | K. Bong Chan | KOR | L. Thamae | LES |
| | 2 | S. Tavakoli | IRI | N. Kloske | GER |
| | 3 | M. Saddiek | KSA | N. Sostaric | CRO |
| | | G. Agapiou | CYP | A. Mavletkulova | RUS |
| Bantam | 1 | M. Beback Asl | IRI | S. Noskova | RUS |
| | 2 | M. Ba`atta | KSA | C. Savioli | ITA |
| | 3 | M. Komane | LES | J. Eun Joo | KOR |
| | | C. Ignatiou | CYP | | |
| Feather | 1 | C. Jin Ho | KOR | M. Karpathaki | GRE |
| | 2 | G. Lo Pinto | ITA | D. Creti | GER |
| | 3 | V. Abdolahie | IRI | E. Aseeva | RUS |
| | | I. Radojcic | CRO | Z. Asoli | CRO |
| Light | 1 | A. Tajeek | IRI | **E. Evans** | **USA** |
| | 2 | K. Taek Jong | KOR | J. Jung Suk | KOR |
| | 3 | A. Gainiakhmetov | RUS | E. Charekovskaya | RUS |
| | | P. Soflanos | GRE | M. Lijane | LES |
| Welter | 1 | A. Hong Youb | KOR | M. Drosidou | GRE |
| | 2 | M. De Meo | ITA | K. Chayer | CAN |
| | 3 | T. Mzini | LES | **A. Williams** | **USA** |
| | | R. Ashori | IRI | P. Seon Suk | KOR |
| Middle | 1 | M. Scheiterbauer | GER | A. Girg | GER |
| | 2 | M. Mokhosi | LES | R. Crkvenac | CRO |
| | 3 | F. Aslani Ghiyase | IRI | A. Chayer | CAN |
| | | A. Lebedev | RUS | **C. Bales** | **USA** |
| Heavy | 1 | K. Jung Kyu | KOR | N. Ivanova | RUS |
| | 2 | M. Nitschke | GER | J. Ibrahimovic | CRO |
| | 3 | M. Annatcheh | IRI | **A. Conley** | **USA** |
| | | M. Shoustov | RUS | A. Greb | GER |

# CHAPTER 14
# BENNETT AND MEDINA STEP UP AS COACHES

1998 All-Army Taekwondo Team

The 1998 season began with a series of miscommunications; these miscommunications led to the All-Army trial camp and team selection being moved from its home at Fort Indiantown Gap, Pennsylvania to Fort Carson, Colorado. How and why this happened? No one really knows all the details, so I won't speculate on what happened. Suffice it to say that the program was returned to its original home at The Gap. 1997 All-Army Heavyweight and national medalist Michael Bennett was selected as this year's head coach with Rafael Medina continuing as assistant coach. Although this year's team was slightly smaller than the '97 team, it was still a very talented ensemble of returning Army, Armed Forces, and national medalists and champions.

There were two new additions to the team: a young man from Hawaii named Bobbie (no one remembers his last name), Eric Laurin, and Hunter Samuels. The veteran players returning to the team were John Swan, formerly of the Fort Hood Taekwondo Team from 1996 to 1998, Kevin Williams, Darryl Woods, Todd Angel, Andrew Roberts, Alicia Williams, and Tricia Demorath.

For undisclosed reasons, Coach Bennett stepped down as head coach of the Army team, and Coach Rafael Medina assumed the duties as All-Army and Armed Forces Taekwondo Team Head Coach. Medina's knowledge and abilities as a coach would soon be put to the test at the 10th World Military Championships. What made this year's competition extremely important was the fact that it would be held on US soil at Fort Hood, Texas, in late November 1998.

At the end of the competition, the US Armed Forces team won a total of six medals with Kevin Jones of the US Air Force winning bronze in the Men's Featherweight division, Andrew "Drew" Roberts winning silver in the Men's Lightweight division, Paul Nelson takes the gold in the men's Welterweight Division, and another US Armed Forces player named Remington also takes the bronze in the Men's Welterweight division. Eric Laurin secured a bronze, rounding off the men's team, and lastly Elizabeth Evans takes gold in the Women's Featherweight division. Indeed, it was the US Armed Forces Team's finest hour. Coach Medina and the team overall had much to be proud of that day.

Team USA: 10th CISM Military Taekwondo Championship 1998, Fort Hood, TX

1998 Armed Forces Taekwondo Team

1998 Armed Forces Taekwondo Team in their dress uniforms

While the Armed Forces team celebrated their victory at Fort Hood, another All-Army Taekwondo athlete was blazing a trail of his own in Germany: Louis Davis (Lol, who did you think I was gonna mention!?). Against the advice of his fellow Fort Hood Taekwondo teammates, who strongly suggested that he get assigned to Korea in order to train with Coach Bobby Clayton, Louis chose to be reassigned to Kitzingen, Germany.

While assigned to the 12th Chemical Company, he began searching for a local Taekwondo club that would help him maintain and increase his growing Taekwondo skill set. His first stop led him to a small Taekwondo club outside Harvey Barracks. While training at this small club, he came across a magazine called *Taekwondo Aktuell*.

With the help of his new girlfriend, a local German named Katerina, Louis quickly learned of the Bayern Taekwondo circuit and began preparing to compete within it. By happenstance, Louis came into contact with a Taekwondo club in Würzburg called TGW; the club was headed by Master Peter Müller.

By early fall of 1998, Louis entered the German Open and got his first taste of international competition. Following in the footsteps of the former Fort Bragg Taekwondo team, Louis competed with no coach or fellow teammates behind him, only his limited knowledge of the Taekwondo game and his drive and determination.

Louis did, however, receive guidance and instruction by phone directly from Coach Rafael Medina. Coach Medina instructed him on how to manage his time before his first formation, during lunchtime and on the weekends. Louis followed Coach Medina's instructions to the letter and followed the instruction received from Peter Müller at night when he trained at the TGW club.

The hard work paid off. By late October, Louis entered a regional championship held at Bad Winsheim, Germany. Competing in the welterweight division, Louis effortlessly won the gold medal. Two months later he took gold again at the Bayern Championships in Garmisch-Partenkirchen, Germany.

Louis' performance at Bad Winsheim gained the attention of Grandmaster Bak Nam Cho of Cho's Tiger Academy, from whom Louis would later receive support at Garmisch-Partenkirchen. Louis would also be invited to compete in Italy's first Italian Open, held in Naples, Italy in December of that same year.

However, that story is for another time...

Louis Davis at the 2000 Bavarian Championships in Stuttgart, Germany

# CHAPTER 15
# INTERVIEW WITH COACH BOBBY CLAYTON

Bobby Clayton at the Foreigners Taekwondo Championship at Kukkiwon in the early 1980s

**Louis Davis**: Coach Clayton, I want to thank you for taking the time to sit down and talk about your time as both competitor and head coach for the Armed Forces Taekwondo Team. So, my first question: when did you first begin Taekwondo training, and what drew you into it?

**Bobby Clayton**: Oh, it goes way back. As an Amerasian growing up in Korea, being both African American and Korean and I got picked on a lot and. This led me to become interested in Taekwondo. So, I asked my mom. I was like, "Hey, I want to take Taekwondo." And I remember her telling me, "Once you start, there's no quitting." I think that she thought that eventually I would need Taekwondo to defend myself against being bullied. When you grow up in a society where being half Korean, wasn't in the mainstream, you were considered as an outcast.

My training initially began in 1969 and earned my black belt either in 1970 or '71. Back then, we had two places where we worked out on the US base gym (Which was then known as called RC-4) Which translates as Recreational Center Four. Prior to training at a gym on a US installation we used the gym in the downtown (aka) the village gym which was a US Army GP Medium tent with rice-straw mats, not like ones used in a Japanese dojo, these were the rice-straw bags that were used by Koreans to put rice in them; these bags were cut in half and spread on the ground throughout the tent. This was the tent that we worked out in. I remember training in this particular tent very well because whenever it rained, the tent leaked.

That part of those early days that I can clearly remember, even after all these years. That's how I got involved with Taekwondo back in Korea. We were located close to an area called the JSA, which is a joint security area which is close to the DMZ.

**Louis Davis**: Wow, you were really close to the line.

**Bobby Clayton**: Let me show you something. I got some photos I'd like you to see. (*He shows me a few pictures saved to his cellphone.*) These pictures were taken while I was at Kukkiwon competing in a tournament called the Foreigners Championship. I believe I was either 19 or 20 years old when I won at this event. Do you remember what I told you about the Foreigners Championship?

**Louis Davis**: Yeah, you mentioned it the last time we spoke.

**Bobby Clayton**: Back then, a women's championship didn't exist in Korea, so the Koreans held both the Foreigners Championship and a women's championship were held together. You'll see that both the foreigners and women were on this side. Because I spoke Korean, I would often be asked to make the declaration of sportsmanship to Dr. Un Yong Kim to begin the event. This one was also taken after the tournament. And that guy right there, this guy, he eventually went on to become world champion Egyptian Middleweight. Amir… (*Coach Clayton wasn't able to recall Amir's last name.*)

**Louis Davis**: I heard about him! For the purpose of the book, could you send me these photos?

**Bobby Clayton**: Back then, The Foreigners Championship had some high-class guys competing in it. The Korean government also sponsored some coaching academy courses that I took. These courses required you to take an examination in order to be certified. So, during the examination, you would be required to wear your name and your student number on the left side of your uniform. This is exactly how I took the certification test at Kukkiwon. I believe that I was one of the first non-Korean citizens to ever go through the Korean instructor certification course.

**Louis Davis**: So, growing up in Korea, you mentioned that you were bullied, and you went into Taekwondo. During your training, how often did you experience racism in the dojang?

Bobby Clayton while attending Robert G. Cole High School in Fort Sam Houston, San Antonio, TX. He was a senior when this picture was taken.

**Bobby Clayton**: I didn't experience it in the dojang, but I definitely experienced it outside of the dojang. It went on for quite some time, actually. That's rather a sensitive area to go into because it's been on TV and everywhere.

**Louis Davis**: I quite understand. My next question; When did you enlist in the Army? And what was the deciding factor for enlisting?

**Bobby Clayton**: You're not going to believe this. It was early 1980. There was a program called Early Delayed Entry Program.

**Louis Davis**: The delayed entry program, eh? Nice!

**Bobby Clayton**: Maybe early wasn't a part of it, but I was still 17 when I graduated, and I still needed my dad's signature to get into the Army.

Among the few reasons I chose to enlist, one of them was to continue my Taekwondo training, by getting stationed in Korea so that I could continue to learn more. I attended Basic training at Fort Jackson South Carolina and AIT (Advanced Individual Training) at Fort Lee Virginia. Upon completion of my basic and advanced training, I got my wish and I returned to South Korea, my assignment was Camp Humphreys, and I was attached to a maintenance battalion as a supply clerk.

**Louis Davis**: So, you were in Logistics back then?

**Bobby Clayton**: Yeah, I was hoping that I would get assigned to Yongsan, but I assigned to Camp Humphreys instead. It took me quite a while to find the right school to continue my Taekwondo training. Eventually I found one through one of the local Koreans known as a "Houseboy". Have you heard of a system called the houseboy system?

**Louis Davis**: Houseboy system? I'm afraid that I've never heard of this. Care to elaborate?

**Bobby Clayton**: A "houseboy" was term used to describe a Korean gentlemen would take care of your uniform and polish your boots. You would pay the gentleman like $20 a month or $30 a month for their services. They would do the following: shine your boots, clean your room and do your laundry. So hired a houseboy and his services allowed me time which I used to go and train.

The distance from the barracks to the main gate back at Camp Humphreys was about a mile. So, I would run from barracks to the gate, go out and catch a Korean bus, go down to Tonguetek City. And I found the house through the reason why I mentioned houseboy, a Korean gentleman. And I asked him, I said, "Do you know of any good Taekwondo schools?" He said, "Yeah, I know of one." I then asked him, "Can you take me there?"

He's like, "Sure." I went out to Pentagon City, and the instructor's last name was Young Wei Sok; he was retired ROK (Republic of Korea) Special Forces, and he was on the ROK Special Forces team. And it was no, no, no, this was post-Vietnam. And so, I paid for private lessons from Young Wei Sok, but I trained with the group. Young Wei Sok was a fighter; a Jidokwon style fighter. So, I trained and competed under him, and participated in the, first, the Foreigners Championship that was held at Kukkiwon every year.

The Foreigners championship was for non-Korean competitors, you could be Chinese, you could be European, you could be from any city in America, you could be Japanese; anybody could compete in the foreigner's championship as long as you weren't Korean. So that's where I first started the using the competition style of Taekwondo. Then, of course, in the States as I was growing up, because I left Korea in 1977. My family relocated to Aberdeen, proving ground first, then from there we relocated to San Antonio. While at Aberdeen proving ground, my dad introduced me to two soldiers who practiced Taekwondo. They were not used to the competition style, but they practiced Taekwondo.

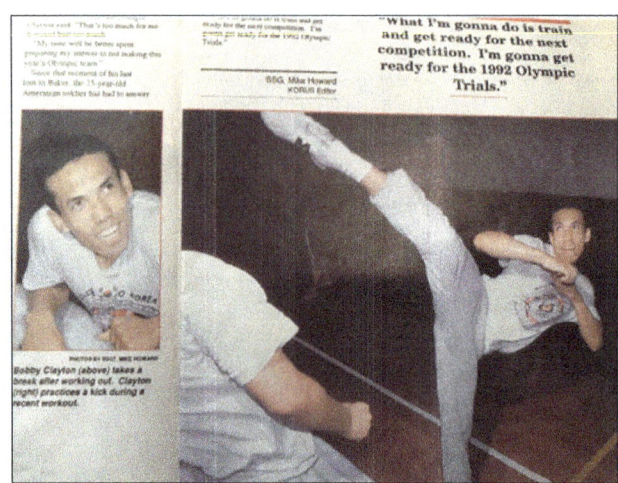

Coach Bobby Clayton in his early days as a competitor

They were initially stationed in Korea. I still remember their names, two black gentlemen. One of two gentlemen was named Ed Brown who was a Master of Japanese Goju Ryu Karate, the other gentleman was named Freddie Gidden, the style he practiced was called Shim Do. It wasn't Taekwondo, but it was called Shim Do.

**Louis Davis**: Ed Brown and Freddie Gidden? These guys were members of the Black Karate Federation!

**Bobby Clayton**: I think so, both men were well known back then and while they were stationed in Korea at that time, I trained under them. Then, of course, after eight months we parted ways. While I was still a high school student as a 10th or 11th grader, my father got assigned to Fort Sam Houston. During our time there we learned of a couple of Korean instructors who were teaching martial arts at main post gym, before that, there were just a bunch of soldiers who took Taekwondo or took Karate while stationed in Japan. We would get together weeknights and weekends and we would spar with one another. I often fought with these men and all of our matches were off record.

**Louis Davis**: Which reminds me, I wanted to ask you about your time training at Dongsan High School. How does fit Dongsan High School into the picture? When did you begin training with them? Was it before or after you enlisted?

**Bobby Clayton**: It was after I enlisted, and I was assigned to Camp Humphreys. I was only able to stay one year and my request for an extension was denied. So, I went back to the States, and was assigned to Fort Carson, Colorado. That's where I met Donald Jackson and there was another guy named Curtis Powell, who is now a Korean instructor at the Defense Language Institute in Monterey California. Curtis competed as a Heavyweight in the sport.

So, there was a group of people who came together and trained together in competition style, Taekwondo. And then of course, there was a school in Colorado Springs, Nam's Taekwondo. Remember the picture that I showed you? Nam Te Hi has a son who was running a school in Colorado Springs at that time. So, I trained there whenever I could.

Back then, I was a 20-year-old, E4 who didn't have a car. So, I'd usually get a ride from older service members who were also involved with martial arts, and I would occasionally go out and compete and train at that school. After one year at Fort Carson, I was reassigned to Korea which was in 1983 and I was assigned to an area currently known as K16. Back then, the village surrounding the K16 area wasn't part of the city, so there weren't any Taekwondo nearby schools. I did some research and found a school in the city of Sonna.

**Louis Davis**: How were you able to conduct your research? Did you use the *Taekwondo Times* magazine as a reference?

**Bobby Clayton**: No, it was the Korean version similar to the Taekwondo Times magazine. So, I looked there, and I look at the competitions; high school competition, university competition, first place, same school, some of the top high schools, I was readily able to recognize which one was top, not school. So, the nearest one was in Songnau, which is not too far from K16, Pung Zhang High School. I went there once or twice, had a little minor, minor...

**Louis Davis**: Disagreement? Let's call it a disagreement.

**Bobby Clayton**: It wasn't even disagreement. I'll tell you, like, if someone tried to pull my wallet out of my pockets, I was like, this is not the place I would want to come to. So, I looked around for a school and then found this gym called Hangu Chukwa. It's a Korean gym. That was the headquarters gym for the Jidokwon system. The headmaster of that school, he's the Grandmaster of all the grandmaster, Hi Chongu, who now passed away a couple of years ago, three years ago. And so, it was a long travel from K16 to downtown Seoul, where the school was located, Kanbi Chipun was located.

Usually, I would catch a bus and it would take anywhere from an hour to hour and a half to arrive at my destination. I went there whenever I could, and I would return home no later than twelve midnight. While in the process of training, I learned from the head instructor there that he regularly went to Dongsan High School to conduct promotion tests. So, knowing where Dongsan High School stood ranking-wise, I knew there were a lot of famous fighters who came from there. So, I spoke with some of the instructors, and I let them know that I was interested in competition training.

I remember asking them (In Hangul), "Can you tell me where Dongsan High School is?" I really wanted to go train spar with some of high school kids attending this school. The head instructor gave me permission to go to this famous high school and that's when I met Coach Kim Sei Hyeok; I introduced myself, and politely asked him for permission to train and spar with the other students; he readily said yes to my request. This was the beginning of my training at Dongsan high school, I couldn't train there regularly, but I went whenever I had a day off from work, or if I was able to get off work early, I would then make my way over to Dongsan High School and I would train. This included training on Saturdays as well. This went on from, I believe, 1984 to 1985.

**Louis Davis**: How many hours did you put into each training session at Dongsan Highschool?

**Bobby Clayton**: Training is about two hours, counting warm-ups and just kicking drills. Afterwards I would kick paddles. Now, I had regular training partners which were usually high school students ranging from junior to senior students. Usually, it was a senior student. We would hold paddles for one another and train afterwards. So, in total it was usually four hours; two hours of group training and two hours of individual training. This was the regimen whenever I could to make it to Dongsan High School.

*(The hard work that Bobby Clayton put in while training at Dongsan High School bore some serious fruit. While competing in the lightweight division he won the 8th Army Championships in 1984, and easily winning the competition. However, I later learned that this was just the beginning of his journey.)*

**Louis Davis**: Moving forward, here, how did you first learn about the All-Army Team? I know we spoke about it briefly on the way here. How did you find out about the program? And how did Baldwin, Kubotsu, and everyone that pioneered the Armed Forces Association find out about you?

**Bobby Clayton**: Well, '85 was when I started going to Dongsan High School and started training there, at that time there was a televised world championship being held in Seoul, and I watched it. After watching the event, I realized that the weight category that I was watching was featherweight and lightweight, I think the featherweight and lightweight championship matches that I was watching featured Chris Spence and Greg Baker.

The very next year, in 1986, they switched divisions. Chris Spence went down to Featherweight and Greg Baker fought in the Lightweight division.

Bobby Clayton won the 8th Army Taekwondo Championship in 1984. The late Major General Henry Doctor Jr. was the 2ID G-1 (Commanding General)

When I saw this, I was like, "Wait a minute. If those guys are the current US National Champions (comparing myself with them)," I thought, "I would have a pretty good chance." I quickly realized that if I wanted to make the USA team at that time, there was no way I could do that on my own. I didn't have the means, nor did I know how people became national champions. One thing I DID know, in order to become a national champion, I had to return to the United States.

So rather than extending, because 1986, I was on my second extension, I left Korea in 1985 and was assigned to Fort Bliss, El Paso, Texas. When I arrived at El Paso, I looked at different schools and was fortunate enough to find a Taekwondo school. That's where I met Javier Martinez, and we started training together. I believe in springtime, we all went to the Texas state Championships and if I recall correctly, either in Houston or Dallas... I think that maybe it was in Dallas. We both competed and were selected as members of the Texas state team and went on to Dayton, Ohio for the '86 USTU National Championships, fighting as members of the Texas State team. Back then, they didn't have an official Army team. There *was* a team from Fort Bragg competing at the same time we were competing at the 1986 nationals. I remember one of the members of the Bragg team was this Puerto Rican guy (Rafael Medina). He approached me, starts speaking to me in Spanish. I was like, "Oh, hey, I don't speak the language."

*(One interesting side note: back in his younger days, Bobby Clayton could easily pass for being of Hispanic or Latino heritage and was often greeted by individuals who spoke Spanish.)*

In the process of having that short conversation, either I or Medina asks first, "Are you in the Army?" So, we both found out that we were in the Army, and I learned that there was a team, one team at Fort Bragg, but not an official team that was sponsored by the Army or All-Army Sports.

**Louis Davis**: As a matter of fact, those guys were sponsored by DA (Department of the Army). To go to the 1985 US National Championships in Hartford, Connecticut, they received sponsorship because of what happened in North Carolina, and Pedro Laboy was coaching them at that time, and they were coaching each other. Laboy's knowledge of the sport was how they won the North Carolina state championships. It was with the help of Grandmaster Myong Mayes; they got the much-needed Support they needed to go to the event.

I understand that you placed within the top four while competing in the 1986 National Championships in Dayton, Ohio; is that correct?

**Bobby Clayton**: Yes, I placed second that year, and first place went to Kareem Ali Jabar and third went to Greg Baker, whom I saw on TV in 1985 competing as a Featherweight, who fought Lightweight in 1986, and I beat him on his own home turf.

**Louis Davis**: What was the score?

**Bobby Clayton**: Well…it was clear enough where there was no doubt *who* won the fight. That's when I guess the US National Taekwondo Championship first got introduced to "cut kick." And the term, I didn't know this at the time, but the term "naraban" translated to English means jump spinning roundhouse kick; I was introduced to this term by Mr. Steven Capener.

He said, "Yeah, don't you remember? You came to 1986 Nationals?" And that's when I first met him. And as I was describing the kick, I told him it was called naraban. So thus, the beginning of the usage of the term naraban by people here in the United States, BUT according to Mr. Capener, it was started by me. I'm the one who imported the term. But in terms of now, that was the year that maybe they have seen the different versions of this kick.

But with the term in the context of fighting in connection with the follow-up techniques, I can comfortably say I think well, ( I don't want to say.) *and* (I could be wrong), that it was me who started with cut kick combinations and cut kick game: broader perspective and countering, some spinning back kick or spinning back kick, spinning kick with a cut kick, following up, that sort of thing.

**Louis Davis**: So, you pioneered the concept of the safe game?

**Bobby Clayton**: Yes, I got it from Dongsan High School. So, after '86, I left El Paso in an effort to pursue further study in Taekwondo, I was reassigned to Korea and went back to Dongsan High School to continue training all the way up until the 1988 Olympic trials.

**Louis Davis**: Let's go back to 1986. Who extended the invitation to you to go to the Armed Forces Trial Camp in Little Creek, Virginia?

**Bobby Clayton**: I don't remember how I found out about the Armed Forces. I remember going to Little Creek for the Armed Forces Trial Camp and ended up getting hurt during the trial camp and I wasn't able to compete in the fight off, and I missed the team, making the team. The injury that I suffered happened while sparring with a heavyweight during the trial camp. That injury ended up having a permanent, long-lasting impact on my hip, leading to a modified version of hip replacement surgery in 2015.

In 1988, I don't know where, but somehow through an official message, I learned about the first All-Army Trial Camp at Fort Indiantown Gap. That was the beginning, the genesis of the All-Army Taekwondo program as part of the All-Army sponsored sports program.

**Louis Davis**: Ah yes, Fort Indiantown Gap; a rundown piece in "No Man's Land," that was perfect for producing champions.

**Bobby Clayton**: It was completely different if you compare the difference between, not to say anything about or highlight myself, but it was a different regimen, a training regimen between the first coach (Bruce Harris) and me. Complete change. Now, here's the thing; complete change in a sense, technical aspect of it. You're right. As you mentioned earlier, after ten miles, morning, ten miles and everything, but the technical session or the Taekwondo session, as a team captain, Coach Harris left that training up to me.

So as an athlete, I designed the training program, the regimen. And as I was going through it, everybody at the camp had gone through it with me. It was a Dongsan style training regimen, but without much explanation because I too had to train and prepare myself while simultaneously help my fellow athletes, Soldier-athletes prepare for competition.

**Louis Davis**: Wow, you wore two hats during that time!

**Bobby Clayton**: Kinda, sort of. This technical regimen that I offered the team continued until I stopped competing in 1992, it went through, as I mentioned earlier, off the record, little incident, little challenge, 1993. Then I got a phone call from All-Army Sports Director Havlicek? I think that's how you pronounce his name. I received this call from him back in 1995 and he asked me if I was interested in the World Class Athlete Program. So, when I signed up for the World Class Athlete Program, Mr. Paul Boltz also recruited me to be the coach. And so that was another good thing. You see, both the All-Army Taekwondo Team and WCAP, while being two separate programs, fall under one organization: Community Family Soldier Support Command, aka CFSC.

**Louis Davis**: Based off a discussion we had earlier, you were definitely the right man for that job.

**Bobby Clayton**: Because I oversaw both programs, I did my best to make sure WCAP athletes were not treated any different than an All-Army athlete. What's the difference? The one's a full-time athlete (WCAP), while the other is a part-time athlete (All-Army). Technically the All-Army athlete is considered extremely part time, like one month out of a year at best, maybe two months out of the year. So, I had to kind of juggle even the training camp at All-Army since WCAP trained under me full time and the All-Army athletes had only one to two months out of a year or two months out of the year to train with me. So, I had to use the WCAP guys to integrate the program/system knowledge and use them to the advantage of the All-Army athletes.

So, I had to mix the two using the All-Army team.

**Louis Davis**: So, WCAP represented the standard that we had to meet as All-Army athletes. Am I correct?

**Bobby Clayton**: Yes, it took me a couple of years. But one at a time. In terms of technique-wise, the All-Army side caught up with WCAP. And that is mainly because there's one person who played a critical role in that. That's a Heavyweight named Michael Bennett.

Bennett took the program and our training regimen program with him back to Fort Hood and established his own full-time program similar in nature to WCAP, so when you guys came back the following year, that helped increase the number of athletes who knew the system/training program. And so as one first year, second year, third year, next thing I know: before, I had to show everything, do this, and do this physically. By 1997 throughout 1999, I didn't have to show them anything. I can say roundhouse kick, lead leg pull, boom-boom-boom. Everybody would just do it. So, the majority was the Soldier-athletes who knew the training content in detail. And then the soldiers who knew little, they were the minority. And so, it was easy for them to follow versus where if you only had four versus thirty who were the minority, who did not know the training regimen or training program in detail.

**Louis Davis**: Bennett's children.

**Bobby Clayton**: Bennett is the one who took the program and basically replicated the program at Fort Hood down at his level and did everything. Not only that; even what I said during the seminar that I would give during the trial camp. You guys memorized the whole thing, so I had to tell the Fort Hood guys to keep quiet because you guys had all the answers. You knew the whole thing, the history, the six structures; you guys were well prepared.

**Louis Davis**: To be honest, all of us Fort Hood guys were trying to impress you.

**Bobby Clayton**: I think I had mentioned that to Bennett once before. Bennett was a critical factor in the Army Taekwondo program, reaching the top status between '96 and 2000.

**Louis Davis**: Bongseok Kim and I have debated this back and forth, but to date Bennett sent two teams from Fort Hood: one in 1996 and the other, 1997. He sent the largest team each year during that time. I still have my first set of TDY orders in a scrapbook that I kept to this very day.

**Bobby Clayton**: Bennett deserves credit for creating a solid grassroots program at the unit level at Fort Hood.

**Louis Davis**: I agree with you on that. He also laid the foundation for the 2ID Taekwondo Team when he trained Johnny Birch Jr.

**Bobby Clayton**: An interesting thing, which has been very valuable overall: the whole overarching program, second program for both All-Army and a contributing factor towards WCAP; one proof factor of Bennett's grassroots program is Andrew Roberts. Drew grew out of Fort Hood's program and then he later became part of WCAP, the other was Kevin Williams (aka Big Dog). One of the things that Bennett did, in addition to ensuring that the detailed aspect of training or technique was passed on to Soldier-athletes, which was a very critical thing, was no matter where he went, he always established a program through which soldiers had ample time, full time in most cases, to train. This was something that he did consistently.

**Louis Davis**: To my knowledge, he duplicated the program that he had set up at Fort Hood, like I said, he laid the foundation for the 2ID team when he trained Johnny Birch Jr. And once he left Korea, he established the Fort Lewis Taekwondo Team at Fort Lewis, Washington. I learned of the Fort Lewis team when I bumped into them and Coach Bennett at the 2003 US Open in Las Vegas.

**Bobby Clayton**: I don't know if anyone has ever given him the credit that he deserves, not at a top level, I think.

**Louis Davis**: I most certainly have. When he got inducted into the Taekwondo Hall of Fame back in 2013, that may have changed things a bit and I think that his efforts were recognized then.

**Bobby Clayton**: I give him that credit. Again, like I said, it was a huge contributing factor in growing Taekwondo in both the Army and the Army World Class Athlete Program. When it was at the latter part of its nascent stage like when I first started, because as I told you previously, there were two different stages of growth in Taekwondo for the Army. Bruce Harris, technically speaking, had a different style of leadership, too. However, with regards to the technical aspect of the All-Army and WCAP Taekwondo program, it was based on what I learned from Dongsan High School, but then it was a mature Bobby or mature, let's say, mature Master Sergeant Clayton, who, yeah, he had to transition himself from the athlete's mindset to the mindset of a coach.

It took a year or two of continuing that evolution of developing myself into a better coach. I could not have done that by myself. Two things became the driving factor for me to think, to improve myself. One was the athletes that I worked with, the other was Rafael Medina. I could not have done this without the support of Rafael Medina. We were a great team, and he is an excellent partner.

**Louis Davis**: Of everyone that we've talked about so far, what made you decide to choose him as your assistant? As your assistant coach?

**Bobby Clayton**: 1995 was my first year as a coach, and if I recall correctly, there weren't many of the original "old school" All-Army Taekwondo team members from the 1988, 1989 era except for Medina. I think Medina was the only one. He's the only one left that came back to compete. If I recall correctly, he may have come back because he'd heard that I was coaching. So, when he came back to the camp and he fought, he was much older, than the other bantamweight guys in his division, however his wisdom and experience was there. Unfortunately, he lost during the fight-offs that year. Later Mr. Boltz came to me, and said, "if you want, you're authorized to have an assistant coach." So, my reaction was "Oh really, Sir?" His response was "Why would I tell you something that you already know?" You know how sometimes he can be?

Mr. Boltz then asked me if I had anyone in mind and my reply was yes, I do, and I think he guessed it right. So, in some way, I think he sensed and saw the same thing that I saw, the old guy with a lot of experience, wisdom, and leadership, too, because I think either he was staff sergeant, or he was sergeant first class. We were the same rank when we coached the team together. And so, when Mr. Boltz gave me the good news, there was no doubt in my mind. I didn't even have to think twice; I knew that I had the perfect candidate. He jokingly asked me, let me guess who that might be and I'm certain that he already knew. Without hesitation Mr. Boltz agreed and that was the beginning of our partnership—the partnership between Bobby Clayton and Rafael Medina.

Bobby Clayton won an open tournament in Colorado Springs, CO, in 1982. Late Great Grandmaster Nam Tae-Hi is presenting an award. GM was the President of the US Taekwondo Federation.

*This partnership, in many eyes to include my own, is often viewed as the perfect storm. Both of these former Army and Armed Forces athletes were exactly what was needed for the program to evolve, more importantly, for the athletes themselves to evolve. In this next section of the book, I will touch on how these two created what Coach Clayton refers to as a paradigm shift for the All-Army and Armed Forces team as the '90s drew to a close.*

*During my interview, I was able to gain a better understanding of Coach Clayton and Coach Medina's mission to build this new generation of athletes which include myself and a host of others. The interview alone is enough to write a book in its own right. There was one thing Bobby Clayton said that stands out. He said that his athletes ARE his credentials. This extends to every All-Army, Armed Forces, WCAP and US National Champion: Rafael Medina, Michael Bennett, Bongseok Kim, and many former All-Army athletes that went on to succeed. These two carried their knowledge forward. WE are their credentials.*

# CHAPTER 16
# THE RETURN OF BOBBY CLAYTON AND RAFAEL MEDINA

1999 All-Army Taekwondo Team, Fort Indiantown Gap, PA

The 1999 season marked another turning point for many of the athletes who would return to Fort Indiantown Gap for that season. For some it was in preparation for the Pan-American Games, for others it was to settle old rivalries, and for others new to the team it was simply an opportunity to prove themselves worthy of being selected as a member of the team.

One of the standout newcomers was a highly skilled young man named David Bartlett. At that time, Bartlett had an impressive string of wins as a junior prior to enlisting and would go on to become the Army's newest Featherweight.

Current members of the Army World Class Athlete Program Andrew Roberts and Kevin Williams would return, competing in the Lightweight and Heavyweight divisions, 1998 Army Team members Eric Laurin and Todd Angel would return to vie for the Middleweight slot, 1997 Army Team member Nicolau Andradae returned to compete for the Featherweight slot, and former 1997 Army Team members Darryl Woods and Louis Davis would return to settle an old rivalry from the 1997 Armed Forces Championships.

Two weeks later it was once again time for the fight-off for positions on the 1999 All-Army Taekwondo Team. In certain weight classes, the fight-off would be hotly contested as each of the following divisions had at least one or more members of past All-Army and Armed Forces Taekwondo teams:

## ONE TEAM ONE FIGHT ONE FAMILY

Lightweight, Welterweight, Middleweight and Heavyweight:

In the Featherweight division finals, Bartlett was matched against Jean Luc Perigone. Although Perigone fought valiantly, he was easily outclassed and nearly knocked out by Bartlett. In the Lightweight division, Andrew Roberts unleashed his arsenal of fast, explosive, and powerful kicks which would, upon impact, echo throughout the gym at Blue Mountain Sports Arena.

Eric Laurin would defeat former 1998 Army Teammate Todd Angel in a very close match. In the Heavyweight division, Kevin Williams would easily defeat Aaron Andrews, capturing the top spot in the division; in the Men's Welterweight division, Darryl Woods was in top form defeating his first opponent, future 2001 All-Army Taekwondo team member Eugene LaRocca. Louis Davis would soundly defeat his opponent capitalizing on his speed and natural athleticism.

Woods and Davis would meet in the finals; the match would begin initially as a potential stalemate until Davis's kicks began to find their mark. By the second round it was clear that Davis was the dominant competitor and at the end of the match Davis was declared the winner.

The 1999 All-Army Team consisted of the following:

**Men's Team**
Flyweight: Jin S. Im
Bantamweight: Ashley Serrano
Featherweight: David Bartlett
Featherweight: Daniel Clifford
Lightweight: Andrew Roberts
Welterweight: Louis Davis
Middleweight: Eric Laurin
Heavyweight: Kevin Williams
Women's Team
Finweight: Jennifer Anderson
Flyweight: Tricia Demorath

One of the customary pranks that the team would play on one another revolved around the infamous ice bucket. The team would dump a bucket of ice on one of our unsuspecting teammates while he was taking a shower. Try to imagine being in the middle of a hot shower and suddenly on the receiving end of a brisk bucket of ice! That mix of hot-n-cold is indescribable, but when you're the one playing the prank, the laughs would be without end!

The bonds formed by this unusual, yet talented group of men and women would stand the test of time, as many of us who "stepped up and stepped out" for this team maintain those strong bonds to this day. This is the other meaning of our shared motto "One Team One Fight One Family." Our days in uniform have long since passed and yet we still refer to each other as brothers (and sisters).

Several former athletes that perfectly illustrate this sense of brotherhood would be the relationship between Paul Nelson, Kevin Williams, and Andrew Roberts. During the 1999 season, these men seemed inseparable.

**The Infamous Gray ALL-Army Vans**

During our years at The Gap, All-Army Sports provided the team with transportation via two gray, late-model GMC vans. Their intended purpose was to transport us to and from the Harrisburg airport, the gym, Mr. Boltz's office, the storage room (to pick up the competition mats and other Taekwondo equipment), the grocery store (for those who were cutting weight), occasionally, trips into Harrisburg to go to the mall, small field trips to promote the team, and for trips to Canada when we participated in the North American Friendship Games with Canada.

In short, these two vehicles were the unsung heroes of the Armed Forces Taekwondo program. And I could not in good conscience forget about these two workhorses. In the pictures below is one of the last remaining vans.

(Top) Coach Bongsok Kim poses with the infamous grey van #1
(Bottom) Louis Davis and GM Rafael Medina pose in front of grey van #

### Drew's Last National Championship

The 1999-Armed Forces Championships ended with the Army team completely dominating the competition. This year the Army occupied nearly all of the slots available for the Armed Forces team. The 25th USTU National Championships were held at Emory Riddle University in Daytona Beach, Florida. The Armed Forces Team was again poised to make history.

Things got off to a shaky start initially due to most of the team getting a slight case of sunburn while hangin' out at Daytona Beach. The team would be reunited with former All-Army Lightweight champion Reginald Perry, who flew to Florida in support of the team, and his former teammates Coach Clayton and Coach Medina.

All eyes were seemingly focused on Andrew Roberts, as he battled his way through the competition. Very few people knew that Drew had asthma, yet despite this handicap (if you'd call it that), Drew dominated nearly every match leading up to the semifinals.

Under Coach Clayton's guidance, Drew took advantage of opponents who waited a little too long to attack and he would always be three steps ahead of his opponents.

In the finals, Drew unleashed his signature kicking combination with all four kicks landing solid and precise. During the final seconds of the match, it was clear that Drew had successfully become a national champion. I remember witnessing him jump for joy, and seconds later with this calm, cool expression, he'd turn and face the team, shrug his shoulders, saying to us "What?"

The team, overall, won a total of four medals this year: two gold medals, one silver medal and one bronze medal, and Coach Clayton was awarded Coach of the Year for the team's incredible performance. Andrew Roberts would be remembered as the only Army and Armed Forces athlete who successfully beat the best in the country as an asthmatic.

### The World Military Games

With Sydney being selected as the site of the 2000 Summer Olympic Games, the competition for the privilege of representing one's nation remained very steep. The US Armed Forces Taekwondo Team knew the stakes and were confident that they would successfully place one of their members on the upcoming 2000 US Olympic Taekwondo Team.

Coach Medina assumed the position as head coach of the Armed Forces Taekwondo Team with the following Armed Forces champions: for the men's team, all from the US Army, Featherweight David Bartlett, Lightweight Andrew Roberts, Middleweight Eric Laurin, and Heavyweight Kevin Williams.

The women's team consisted of Flyweight Tricia Demorath representing the US Army, Lightweight Elizabeth Evans representing the US Navy, Welterweight Rachel Ridenour representing the US Army, and Heavyweight Jennifer Warf also representing the US Army.

During the training camp, a last-minute addition to the team arrived in the form of returning CISM gold medalist Paul Nelson replacing Louis Davis in the Welterweight division. Louis was, however, allowed to travel with the team as an alternate should Nelson be unable to compete.

The team departed Fort Indiantown Gap, heading to Aberdeen Proving Grounds in Maryland to receive their official warm-up uniforms and additional gear such as bags, T-shirts, polo shirts and other goodies that the athletes could trade with other nations. Linking up with other US Armed Forces athletes from different sports, flying out of Baltimore Airport, they departed for Croatia, confident yet ready for the task ahead. After a very long flight, the team finally arrived at Zagreb for the 2nd World Military Games.

I vividly remember our arrival in Croatia; we left our plane and boarded a bus which took us to the terminal. The welcome that we received will be forever burned into my memory. I was one of the first US service members to walk through the door and what greeted us was a pleasant surprise!

Standing on both sides of a long, red carpet were dozens of pretty girls dressed in drum majorette uniforms! I remember thinking "OH DAYUM!!!!!" as I quickly grabbed my camera and snapped as many pictures of this welcome as I could. Unfortunately, much to my disappointment, *none* of those shots turned out right.

During the welcome ceremony, the team was greeted by the US Ambassador living in Croatia at that time, who came to show his support of the games and the US Armed Forces athletes attending the event. After a day or two of getting acclimated and fighting off the jet lag, the team held one final training session to shake off the remaining traces of jet lag and stay loose before the big day.

The official opening ceremony of the 1999 World Military Games was held at Maksimir Stadium in Zagreb, with live broadcasts of the event on Croation national television. The event itself mirrored the official Summer Olympic games in every detail except the contestants were in military uniform. What very few people know is that many of these military athletes were also members of their country's Olympic team.

There were a total of 6,734 athletes representing 82 different countries throughout the world in 20 different sports which included: Track and Field (including marathon), Basketball, Boxing, Cycling, Fencing, Football/Soccer, Handball, Judo, Orienteering, Wrestling, Swimming (water-polo, diving, lifesaving), Parachuting, Military Pentathlon, Naval Pentathlon, Taekwondo, Shooting, Triathlon, Volleyball and two demonstration sports : Rowing and Canoe-kayak. This is an example of just how high the stakes were for the 2nd World Military Games this time around.

At the end of the competition, the US Armed Forces Taekwondo Team brought home two bronze medals, one belonging to Elizabeth Evans, the other belonging to Andrew Roberts. Sadly, this would be the last world event that Andrew Roberts would attend as a member of our team.

1999-Armed Forces Taekwondo Team in Zagreb, Croatia

# 11th World Military Taekwondo Championships (CISM)

Date: August 11-14, 1999
Place: Karlovac Spots Hall, Zagreb, Croatia

| Participating Nations | Male - 21 (Albania, Austria, Azerbaijan, Belgium, Canada, china, Croatia, Cyprus, France, Georgia, Germany, Greece, Italy, Kenya, Korea, Latvia, Lesotho, Netherlands, Saudi Arabia, USA, Vietnam<br>Female - 11 (Canada, China, Croatia, Germany, Greece, Italy, Kenya, Korea, Latvia, Lesotho, USA)<br>Total - 21 |
|---|---|

| Weight | Rank | Male | | Female | |
|---|---|---|---|---|---|
| | | Name | Nationality | Name | Nationality |
| Fin | 1 | Dong-Jong PARK | KOR | Fantao Kong | CHN |
| | 2 | Ghaleb Al Shamrani | KSA | M. Thandi Lemphane | LES |
| | 3 | Benedic Lebohang Ntsitsi | LES | Michaela Wegner | GER |
| | | Elnur Amanov | AZE | | |
| Fly | 1 | Ludovic Vo | FRA | Nicole Kloske | GER |
| | 2 | Young-Taek YOO | KOR | Alina Likeleli Thamae | LES |
| | 3 | Huasheng Liu | CHN | Eva Politeo | CRO |
| | | Muhammed Baata | KSA | | |
| Bantam | 1 | Soon Tae LEE | KOR | Inga Motmillere | LAT |
| | 2 | Christakis Erakleous | CYP | Ivona Skelin | CRO |
| | 3 | Erol Denk | GER | Su Wang | CHN |
| | | Igor Radojcic | CRO | | |
| Feather | 1 | Young-Jin YOO | KOR | Liyue Zhang | CHN |
| | 2 | Athanassios Balilis | GRE | Marijeta Celic | CRO |
| | 3 | Michael Aloisio | FRA | **Elizabeth Evans** | **USA** |
| | | Dennis Bekkers | NED | Alla Popova | LAT |
| Light | 1 | In-Dong KIM | KOR | Huijing Zhang | CHN |
| | 2 | Bel-Aziz Acharki | GER | Jung-Yun KIM | KOR |
| | 3 | **Andrew Roberts** | **USA** | Dejana Gajic | CRO |
| | | Krunoslav Markulin | CRO | | |
| Welter | 1 | Nam-Woong KIM | KOR | Luming He Hwang | CHN |
| | 2 | Michael Savvas | CYP | Renata Crikvenac | CRO |
| | 3 | Antonino Cutugno | ITA | Morfo Drosidou | GRE |
| | | Fagan Umudov | AZE | | |
| Middle | 1 | Boo-Kwon KIM | KOR | Zhong Chen | CHN |
| | 2 | Marco Scheiterbauer | GER | Iva Gavez | CRO |
| | 3 | Haibin Yu | CHN | Anick Chayer | CAN |
| | | Seif Martin Oduor | KEN | | |
| Heavy | 1 | Yong-Hee CHA | KOR | Natasa Vezmar | CRO |
| | 2 | Hastings Ngala Munai | KEN | P. Alphoncinah Lala | LES |
| | 3 | Antonio Vocale | ITA | | |
| | | Ronnie Van Der Burgt | NED | | |

# CHAPTER 17
# THE END OF AN ERA

1996 All-Army Taekwondo Team poses for the camera.

The 2000 season would best be described as a time of change for the All-Army and Armed Forces Taekwondo Team. With the 2000 Summer Olympic Games on the horizon, there were some adjustments made to the sport by the World Taekwondo Federation, and the IOC for the upcoming debut of Taekwondo as an official sport in the summer games.

Instead of featuring eight weight classes, it was decided to combine Finweight and Flyweight, becoming Olympic Flyweight; Bantamweight and Featherweight became Olympic Featherweight; Lightweight and Welterweight became Olympic Welterweight; and Middleweight and Heavyweight became Olympic Heavyweight. With these changes in mind, the soldiers assigned to the Army

World Class Athlete Program were well prepared for the task at hand.

The 2000 Army Team once again fielded a very strong team that year; it was a mix of returning national medalists such as Kevin Williams, Rachel Ridenour, David Bartlett, and international medalist and Russian national champion Yelena Pisarenko. Also returning were previous Army and Armed Forces Taekwondo Team members Ashley Serrano and Tricia Demorath, and returning to their second team trials were William "Big Willie" Chalmers, Aaron Andrews, and Eugene LaRocca. Former Fort Hood Taekwondo Team member Petra Kaui and newcomer Daryll "Ox" Rydholm would be selected as members of this year's All-Army and Armed Forces Taekwondo Team.

At the 26th USTU National Championships, the team would see the return of former national champion and All-Army Taekwondo alumnae Jada Monroe, who would defeat Petra Kaui in the Women's Middleweight division in the semifinals, and Rachel Ridenour, who would win the bronze medal in the Women's Welterweight division.

## Medina's Time as Coach of WCAP

During this time, Coach Medina assumed the position as head coach of the WCAP program where he continued the training of Andrew Roberts, Kevin Williams, and David Bartlett. Medina made changes to the usual training schedule by taking the athletes to different places in order to increase their experience and exposure to different countries. The following document commemorates the team's activities.

## 26th National Taekwondo Championships
US Air Force Academy
Colorado Springs, CO
May 11-14, 2000

### Fin

**Men**
1st Omar Esposo (CA)
2nd Robert Yoon (TX)
3rd Alvin Marquez (CA)
3rd Anthony Melella (NY)

**Women**
1st Carolyn Shu (CA)
2nd Rachel Marcial (CA)
3rd Heather Cravens (OK)
3rd Rebecca Epting (PA)

### Fly

**Men**
1st Tim Thackrey (CA)
2nd Phil Yoon (OR)
3rd Naveed Ashraf (TX)
3rd Ken Tran (OH)

**Women**
1st Chrissy Adamo (NJ)
2nd Stephanie Bowman (TX)
3rd Shea Hipp (FL)
3rd Kim Kies (NC)

### Bantam

**Men**
1st Jared Gullekson (OK)
2nd Victor Ao (MD)
3rd Samuel Hale (NJ)
3rd Brian Stephen (CO)

**Women**
1st Angela Prescott (FL)
2nd Amitis Pourarian (CA)
3rd Marsha Berry (FL)
3rd Jodi Creech (OK)

### Feather

**Men**
1st Antony Graf (NY)
2nd Steve Rosbarsky (MT)
3rd Jason Alvelais (CA)
3rd Sean Lassak (OH)

**Women**
1st Lynda Laurin (TX)
2nd Sanaz Shahbazi (CA)
3rd Jennifer Huang (NY)
3rd Darcy Kimmich (CO)

### Light

**Men**
1st Tae Kim (IL)
2nd Tom Lynn Jr. (NY)
3rd Christopher Smith (OK)
3rd Jin Suh (NY)

**Women**
1st Elizabeth Mohammed (NY)
2nd Kelly Thorpe (WA)
3rd Andrea Ferkile (FL)
3rd Christa Tomkins (PA)

### Welter

**Men**
1st Jason McEuin (WA)
2nd Clyde Gordon (CA)
3rd David Ibrahim (IL)
3rd Kevin Strantz (CO)

**Women**
1st Bonnie Wiegand (WI)
2nd Charity Maclay (MD)
3rd Tracy Black (NC)
3rd Rachel Ridenour (AY)

### Middle

**Men**
1st Richard An (NH)
2nd Keary Watson (OR)
3rd Eric Laurin (IN)
3rd Eui Y. Lee (MN)

**Women**
1st Sanaz Shahbazi (CA)
2nd Jada Monroe (VA)
3rd Petra Kaui (AY)
3rd Carolyn Stephenson (OH)

### Heavy

**Men**
1st Trent Tompkins (IA)
2nd Charles Alexander (DC)
3rd Stewart Gill (OH)
3rd Michael Tang (NH)

**Women**
1st Miranda Hinrichs (IA)
2nd Christina Bayley (OH)
3rd Carolyn Crowley (MI)
3rd Jamie Hamilton (OH)

CFSC-SF-W                                                                                                17 DEC 1999

MEMORAMDUM THRU Commander, WCAP Detachment

FOR Director, Soldier & Family Support

SUBJECT: Trip/Event After Action Report

Purpose of Trip/Event. To participate in the Korean Tae Kwon Do exchange program (Seoul, Korea)

Discussion of Trip/Event. I arrived in Yongsan, Korea on November 16, 1999 and SGT Williams and PFC Barlett arrived on November 17, 1999. On December 5, 1999 we had team dinner with COL P

Results of Trip/Event. As a result of intense training the athletes were able to defeat the Korean Team. On the first couple of days the following areas were highly improved.

Condition (stamina, strength, etc)
Speed (speed drills, reaction drills)
Techniques (defense and attack)

On the following week the athletes worked out with the Korean team twice a day. They got beat by the Korean team on the first couple of days but after studying their strategies, strengths, and weaknesses the athletes were able to beat every contestant. SGT Williams and PFC Barlett left with the victory and the earned respect of the Korean team. They also went to the Korean National Championships to visualize their strategies, footwork, and where improvements could be made (i.e. how to properly manage the ring). On the last two weeks we stayed focused on reinforcement and corrective training.

Analysis of Trip/Event. The most important thing is that the athletes were able to learn how to adapt to different opponent's techniques/sparing styles. Athletes improved their attack strategy, condition and strength. I gave special thanks to Coach Clayton to settle the training with different Korean teams.

Discussion of Strength and Weaknesses. Working out with the Koreans helped to develop their mental and emotional strength allowing them to gain confidence and concentration for future competitions. Weakness was there is not doubt in my mind that the athletes needed a full time coach to recognize their weaknesses and improve their strengths.

Recommendations. I highly recommend that SGT Roberts and PFC Barlett receive a body fat test at least once every two weeks. Losing to much weight increases the risk of fatigue, loss of strength, dehydration, heat stroke, and even death. This in the best worry athletes monitor their weight health. Also I recommend before that athletes must have windbreakers before coming to Korea if the training is during winter. Korean gyms do not use the heater on less the temperature is below 20 or 30 degrees farinhigh. Windbreakers maintain the body heat and avoid the athletes to catch a cold or pneumonia. Also we should be paid in full per diem due to different times and locations of training. Most of the time Korean teams train from 0700-0800 and from 1300-1800. Even though our stay was on a military installation some athletes need special diets to maintain or monitor their weight. Most of the time food at the dining facility is high in protein and fat. Weight must be monitored before any competition.

Conclusion. There are physically and mentally prepared for the next competition.

Rafael Medina
SFC, USA
Taekwondo Coach

## The Loss of Andrew Roberts: Closing the Ranks

This is a very difficult section to write about because Andrew "Drew" Roberts was more than just a talented lightweight champion. To many of us he was a dear friend and a brother. The last time that I would stand beside him was at the 1999 World Military Games. Drew (to me) always had an upbeat attitude, he'd always find a way to pick you up when you were in a funk, he was a bit of a prankster. If there was any kind of prank or joke being pulled, you could best believe that Drew would be in on it.

Sadly, Drew was taken from us in early 2000 as the result of a cardiac arrest that he suffered and was unable to recover from. That year, the team marched into the USTU Nationals in Colorado Springs with a black band tied around the right arm of their competition doboks in honor of our fallen brother.

I was unable to stand with my teammates during that time due to circumstances that would prevent me from leaving Germany to join the team and to say goodbye my friend and brother. There were many who believed that Andrew would have been the first US Soldier to make the Olympic Team in the sport and possibly medal at the event.

Whenever '90s-era Armed Forces Taekwondo is discussed, you can bet your bottom dollar Drew's name will be mentioned quite a bit. Drew will forever live on in the hearts and minds of his brothers and sisters of the Army and Armed Forces Team.

Rest well, brother. We'll see you on the other side...

## Bobby Clayton and Rafael Medina Retire from the Army: End of an Era

There are sayings, "All good things come to an end" and "Nothing lasts forever." As much as many of us wished that these two incredibly legendary coaches would have remained as the head and assistant coach of the Army and Armed Forces Taekwondo Team, for their own personal reasons the two decided that it was time to pass the torch and to step down.

In doing so, they also decided that it was time to retire from military service together following the 2000 All- Army/Armed Forces Taekwondo season. In an interview with Coach Clayton, he stated that his athletes were his credentials: the athletes he's coached from the time he was captain of the All-Army team until the time he assumed the position of head coach of the Army, Armed Forces and the World Class Athlete Program; all serve as a testament to his abilities, his knowledge and intellect as a coach.

The former All-Army, Armed Forces and WCAP athletes who have carried the torch after these two powerful coaches stepped down, to this very day have passed on the teachings, strategies, training methods, and philosophies that were once taught to them as service-member athletes. These continue to be the credentials of Master Sergeant Bobby Clayton and Master Sergeant Rafael Medina.

I too am a part of this legacy, although (in my humble opinion) I didn't receive nearly as much information as my former coaches and teammates from Fort Hood, a few of which were selected to become part of WCAP or become All-Army Coaches themselves. My achievements as a military athlete still serve to validate the credentials of William Baldwin, Daryl T. Kubotsu, Bruce Harris, Paul J. Boltz, Claudia Berwager, Mike Pizauski, Steve Brown, Phil Cota, Bobby Clayton, Rafael Medina, Pedro Laboy, Mark Green, Luis De La Rosa, Larry Spears, Charles Sexton, Curtis Brown, Bongseok Kim, Reginald Perry, Dwayne Lopp, Carlos Rentas, Edward Givans, Ron Berry, Rachel Ridenour, Schileen Potter, Todd Angel, Howard Clayton, Andrew Roberts, Eric Laurin, Johnny Birch, Gregory Sheppard, Kevin Jones, Freddie McDonald, Brad Carter, Hyun Suk Lee, David Bartlett, Kevin Williams, Missy Cann, Jonathan Fennell, Punarrin Koy, Patrice Remarck, Jody Gibson, Ashley Serrano, Javier Martinez, and a host of others who paved the way for athletes like me, who believed in the program, who encouraged athletes like myself to keep moving forward.

Those who believed in me as an athlete encouraged me to begin this project as a means to preserve our history. To me and to many others, these are more than just names of people who either once served in the US Armed Forces or supported the efforts of the Armed Forces Taekwondo program; they were dear friends, brothers, sisters, coaches, teachers, and mentors. It is something that I am constantly reminded of, especially towards the end of my journey as a competitor.

No matter the outcome of any match that I took part in, I was never alone. They were right there beside me in that ring, whenever I felt tired during training, whenever I experienced any negativity that would derail my training or my efforts in any way, shape or form, they were there in spirit encouraging me to keep moving forward. My opponents weren't simply facing me, they were facing the spirit of several generations of military athletes, all carried by one soldier, at all times. One team stood behind me, one fight at a time, as one family.

# CHAPTER 18
# THE DEFINITION OF A WORLD CLASS ATHLETE

**W**hat is a world class athlete; more importantly what are the expectations of a world class athlete? A world class athlete is defined as an athlete who has achieved the status of "world class" within a given sport that said athlete participates in. Sounds pretty straightforward, doesn't it? I'd be lying to you if I were to say that it is.

The first step to becoming a world class athlete within the realm of Taekwondo (circa 1988 throughout the first decade of the 2000s), one is required to place within the top four in a national or international Taekwondo competition. Within the US there were several events which the US Armed Forces recognized:

- The US Open
- The US National Championships
- The Pan-American Championships

On the international side of house, you had a little more leeway. If the competitor (i.e., the military athlete) were to be stationed abroad, competing in WTF sanctioned and recognized events offered greater opportunities to gain valuable international experience and the opportunity to place within the top four at said international event.

Within the US we have a deep pool of athletes, however, only a select few can be considered "World Class"; within the US Armed Forces that pool is very small.

Within the US Armed Forces regardless of branch of service, the path to become a military world class athlete is far more difficult and possesses an even greater challenge than the challenges posed by our civilian counterpart. Being both a service member *and* an aspiring athlete is a balancing act from hell!

You are required to maintain your military occupational specialty skill set, maintain your basic combat skill set (i.e., weapons training & qualifications), and perform any and all duties above and beyond reproach. You are also required to achieve and maintain your competitive edge and provide proof of said maintenance and proficiency.

Oftentimes one's unit mission operational tempo can and does take precedence above all else; you are constantly reminded that you are a service member FIRST and an athlete second. Should you achieve the level of a Military World Class Athlete, you are expected to carry yourself in a manner that demonstrates esprit de corps at all times, in and out of uniform as well as on and off the Taekwondo competition mats.

One must remember: You are representing the US Armed Forces and the best they have to offer; in doing so you must also continue to not only demonstrate to the world but to your peers and your superiors that you are the best of the best.

This representation is also expressed in your appearance; therefore, you should always maintain a high standard when it comes to your uniform, be it your duty uniform, your dress uniform, even your competition attire.

This high standard will ensure that whenever anyone sees you there is no doubt that you are the best representation of not only WCAP (the World Class Athlete Program) but your respective branch of the US Armed Forces.

There are several points that I will discuss regarding a World Class Athlete:
1. Level of Training
2. Military Appearance
3. Personal Appearance
4. Maintaining the WCAP program
5. The future of the WCAP program
6. What WCAP means to me

**Level of Training**

The level of training for a world class athlete should be different in many ways and said athlete should train harder. Therefore, in order to become the best, you are required to train harder, run farther, faster, and fight harder than *any other* athlete. In addition, you must stay mentally alert, physically strong, and morally correct at all times.

You must be willing to shoulder more than what is expected of you in whatever task, requirement or mission that is presented; you must be willing to give 100 percent and more, bearing in mind that no one in this cruel world will give you anything save almighty God. Through faith, God will grant you the strength, courage, will, inspiration and tools that are vital and will enable you to press forward.

The remainder must come from within, fueled by how badly you really want to succeed. Ask yourself and then answer on a daily basis, "Did I train hard enough today? Did I meet or exceed my expectations?"

**Military & Personal Appearance**

As a military world class athlete, your appearance both on and off duty is crucial since you are under constant observation. Therefore, it is imperative that while on duty, your uniform should be clean, properly pressed, and serviceable. Your grooming and hygiene should also be a direct reflection of the high standards expected of you as a military world class athlete.

The same high standards must be reflected in your appearance while off duty as well, maintaining that high military standard that is expected of any soldier, sailor, airman or marine.

Therefore, your choice of attire should be tasteful regardless of where you might find yourself whilst off duty.

Lastly, a military world class athlete should carry themselves with the utmost respect for themselves, the US Armed Forces regardless of branch of service, and the World Class Athlete Program. Remember, a world class athlete represents the service-member athlete; although you're an athlete, you are still a soldier, sailor, airman or marine. FIRST and FOREMOST.

Carrying oneself as a person of intelligence, wisdom, who is proper and well mannered, you should possess the ability to converse with others, and that your words are clear and concise. This will enable you to engage a group or organization with a clear voice and a confident demeanor.

When the world class athlete appears in front of a group, there should be a sense of pride and honor to be both a US Armed Forces service member *and* a world class athlete. This kind of conduct will distinguish you from others, leaving a lasting impression upon everyone who encounters you. Others should see the standards that I've mentioned and want to become world class themselves. Remember, appearance is what can make or break the person and the image that they are trying to convey.

In exchange for earning the title and position of world class, the military athlete must always present themselves as the best Servicemember-athlete that the US Armed Forces has to offer, and in more ways than one. In closing, draw out people's curiosity by making them ask themselves, "WHY is he/she a world class athlete?" And in doing so let your appearance speak for itself.

**Maintaining the WCAP program**

Maintaining the WCAP program overall is equally as important as being on the program itself. The program is special in many ways; initially, the US Army was the only branch of service that offered this program. Therefore, the burden of maintaining the program rests squarely upon the shoulders of those who are currently in the program, which will ensure that it is available for future generations.

In doing so we will encourage our fellow branches of service to create their own program specifically for their respective branches of service. We must prove and continue to prove that the program is worth having by maintaining high standards and expectations. We must work hard to keep sight of our overall goal, which is to win.

We must also be willing to seek out other up-and-coming competitors (via TIER Missions) with the intent of helping them achieve that same winning spirit that exists within the program. For the program to survive, it must continue to have a deep pool of quality athletes, coaches, and support staff that will contribute 100 percent and then some.

This will ensure that all areas are professionally maintained at all times, which will greatly reflect the hard work, motivation, determination and dedication that contributes to the program's success. My mother once stated that you will only get out of any endeavor WHAT YOU PUT INTO IT! Therefore, if everyone does their part to maintain the program in any and every capacity, it will continue to thrive and flourish for future generations of our Servicemember-athletes to enjoy.

## The Future of the WCAP Program

In order to accomplish this goal, we must continue to put aside any differences that we may have with one another and focus upon our common objective, which is the betterment of our athletes and the survival of the program. It is through the World Class Athlete Program (WCAP) that we can train and produce the world's best athletes within the sport of Taekwondo; it is through this unique military program that we can maintain a team with the winning edge, a team that travels and represents the US Armed Forces worldwide with the pride and distinction that this great nation of ours was built upon. This team will become known for having the following: its ability to win decisively, its spirit, the self-esteem of the athletes and coaching staff, lastly the intelligence and knowledge of the sport.

This team will have the ability to host seminars throughout the US and will have earned the respect of other national teams worldwide. This would be a program for all others to follow and desire to duplicate; therefore, the future relies on those involved, past, present and future.

## What WCAP Means to Me

In closing, this program has meant a lot to me. It has given me an opportunity to become not only a better athlete, but a better soldier, a better leader, a better husband, a better father and overall, a better person. Throughout the years, many have placed great faith in my ability to "be all that I can be" within the sport of Taekwondo. I myself simply wanted to be the best and WCAP offered me the chance to do just that.

My motivation for being on this program stemmed from wanting to be more than just a winner; I wanted to be well respected as an athlete, as a soldier and as a person, to achieve a level of greatness. The program offers select individuals the opportunity to achieve their goals while serving in the Armed Forces, but only a few are up to the challenge.

Coach Bobby Clayton with All-Army Athlete Pedro Cruz-Febo

*One Team. One Fight. One Family.*

One of my most important personal goals is to give my mother a gold medal from either a national or international event. I want to pave the way for other military athletes to be able to have access to the same opportunities that I have been granted access to.

I want this program to continue to grow and flourish for future generations and it is with my hard work that I strongly believe this program *will* be successful. The WCAP program is more than just a personal goal; it's for all who aspire to become a champion and are willing to give that much-needed 100 percent and then some.

*This excerpt of the definition of a world class athlete was written by one of its reigning champions during the 1994 – 1995 Armed Forces Taekwondo season. This athlete knows who he/she is and out of respect for this individual's privacy, they will remain nameless.*

**One Team. One Fight. One Family!**

# CHAPTER 19
# THE GATHERING

One Team, One Fight, One Family!

Prior to my beginning to research the history of The Program, Coach Medina began to host the Armed Forces Alumni reunion, first held in 2015. Coach Medina invited any and all former US Armed Forces Taekwondo athletes to attend, regardless of branch of service. By 2016, the reunion had close to twenty former athletes, champions, coaches, the current president of CISM Taekwondo, and an international referee.

This gathering also gave birth to the US Military Taekwondo Foundation, which would serve as a grassroots organization to identify and assist with preparing the next generation of US Armed Forces Taekwondo athletes. This would serve as the military's own pipeline of talent from its ranks of athletes currently serving in the Armed Forces.

# 2016 ARMED FORCES TAEKWONDO REUNION

**Chief of Mission**

Paul J Boltz
Claudia Berwager
Thomas Allmon

*In Honor of All Coaches*

Aaron A Andrews
Adam Vazquez
Adam Prim
Albert Lee
AJ Bonitati
Alana E Conley
Alex White
Alisha Williams
Andrew Roberts (RIP)
Ashley Richardson
Ashley Serrano
Ascenzo Bonitati
**Brad Carter**
Brandon Shaffer
Bret Moldenhauer
Briot Empeh
**Bruce Harris**
Bryan Lee
**Bobby Clayton**
**Bongseok Kim**
Carlos Mena
Carlos Rentas
Casey McEuin
Charity Beyer
**Charles D Sexton**
Charles Santos
Christopher Sunday
**Curtis Brown**
Daniel Clifford
**Darrell Rydholm**
Daryl Woods
**David Bartlett**
David Ruiz
Deanna Charett

Donald Jackson
**Donovan Rider**
Dwayne Johnson
Dwayne Lopp
Edward Fourquet
Edward Givans
**Elizabeth A Evans**
Eric Hampton
Eric Laurin
Eric Slycord
Frankie Davis
**Freddie McDonald Jr**
Gregory Shepherd
Howard Clayton
Hunter Samuel
**Hyun Suk Lee**
Issac Harris
Ivan Abudo
Jada Monroe
Jaime Houston
**James Arrington**
Jamie Toyota (RIP)
Javier Martinez
Jay Utter
Jean L Perigore
Jennifer Mulgard
Jennifer Warf
Jessica Hope
Jim Im
Jin Gyu Choe
Joanie Guerrero
Jody Gibson
John Reyna

John Swan
Johnny L Birch Jr
**Jonathan Fennell**
Jonathan Scherquest
Joshua Fletcher
Julio Saunders
Kermit Gonzalez
**Kevin D Williams**
Kevin L Jones
Kimmi Boothe
Lamonte Kelly
Larry Spears
Lee SJ Hencshel
London Arevalo
Louis Davis
Louis Torres
**Luis De La Rosa**
Luis Rodda
Maria Juarez
Mark Green
Matthew Dalrymple
Matthew Darby
Melvin Boatner
**Michael Bennett**
Michael Delgado
Michael King
Michael Warner
Mike Kandewen
**Missy Cann**
Nicolau T Andrade
**Patrice Remarck**
Paul Nelson
Pedro Cruz-Febo

Pedro Laboy
Peter Son
Petra Wyche
**Punarin Koy**
Quinton Beach
Rachel M Donaire
**Rachael Ridenour**
**Rafael Medina**
Reginald Perry
Reynaldo Martinez
Rodney Johnson
Ron Berry
**Ron Onyon**
Ryan MacCulloch
Schileen Potter
Scott Potter
Sean Burke
Simmon Behrndt
Steve Ostrander
Steve Whittle
Steven Carter
Taewoo Lee
Tammy Cisneros
Ted Berry
Timothy Grant
Tina Chong
Tricia Demorath
Todd Angel
Troy Evans
**William D Balwin**
William Sexton
William Rider
Willie Chalmers
Yelena Pisarenko
Yong Hun Kim

**Hosted by Rafael Medina**

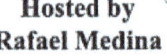

# CLOSING
## IN CONCLUSION

I want to thank you for taking the time to read my work and venture back to times past. Please understand that when I first began to conduct my research, I had very little to go on and assumptions that I once had about the US Armed Forces keeping records of its athletes turned out to be a disappointing falsehood. Then there's the quest to locate former military athletes, teammates, and coaching staff, an even bigger challenge especially if the athletes themselves were no longer active in the sport.

Facebook did in fact help close the gap but there were things such as competition results, team rosters, copies of old travel orders, and other types of memorabilia that would further close the gap on our rich and powerful history from the '80s until 2000.

Google was certainly *not* my friend from time to time and Taekwondo's governing body here in the US did not keep much of the old records and results from the days of the US Taekwondo Union. Lastly, none of us can outrun Father Time; one day death will come for us all.

It is my sincere hope that this book will encourage other former military athletes to come forward with old pictures, or documentation that will continue to bridge the gap. As of 2021, the All-Army Taekwondo program had been cut and is no more. All it would take is the passing of maybe two to three years and all that we have done, all the memories of our shared time together, will fade into obscurity.

Again, thank you from the bottom of my heart.

All Army 1993

The ladies were late

# OUR BIGGEST SUPPORTERS

GRANDMASTER JOHN HOLLOWAY    MASTER JAMES "JOJO" STAGEN    CWO MICKEL CRUZ

Three of the biggest supporters of the US Armed Forces Taekwondo Team were GM John Holloway, Master Jojo Stagen, and Chief Warrant Officer Mickel Cruz. These men have stood by us during some of the most challenging times our team has faced. To this day they still stand in support of our efforts to keep Armed Forces Taekwondo alive. Chief Cruz is currently the only active-duty international referee representing the US and the US Army.

# A TRIBUTE TO GRANDMASTER "HAWK" HAWKINS

# IN HONOR OF
# COACH MISSY CANN, USMC

# EPILOGUE
# CLOSING THOUGHTS

To the current head coach of the Army World Class Athlete Program: Brother, we have seen the hard work and dedication you have put into training this new generation of Armed Forces talent. Make no mistake, Coach Jennings, your efforts have *not* gone unnoticed. At the end of the day, we still got your back if you need us...

Although the program has been cut and Fort Indiantown Gap is no longer our home, our spirit dwells within the hallway of the old barracks that once housed many a military Taekwondo athlete. Our footprints can always be found on the roads that make up our morning running route, and the sounds of athletes kicking targets can still be heard within the walls of Blue Mountain Sports Arena. As the last All-Army and Armed Forces Taekwondo athlete leaves this place, I am reminded that the spirit of the program lives within me...

Mission First

Soldiers Always

# USA ARMED FORCES TAEKWONDO

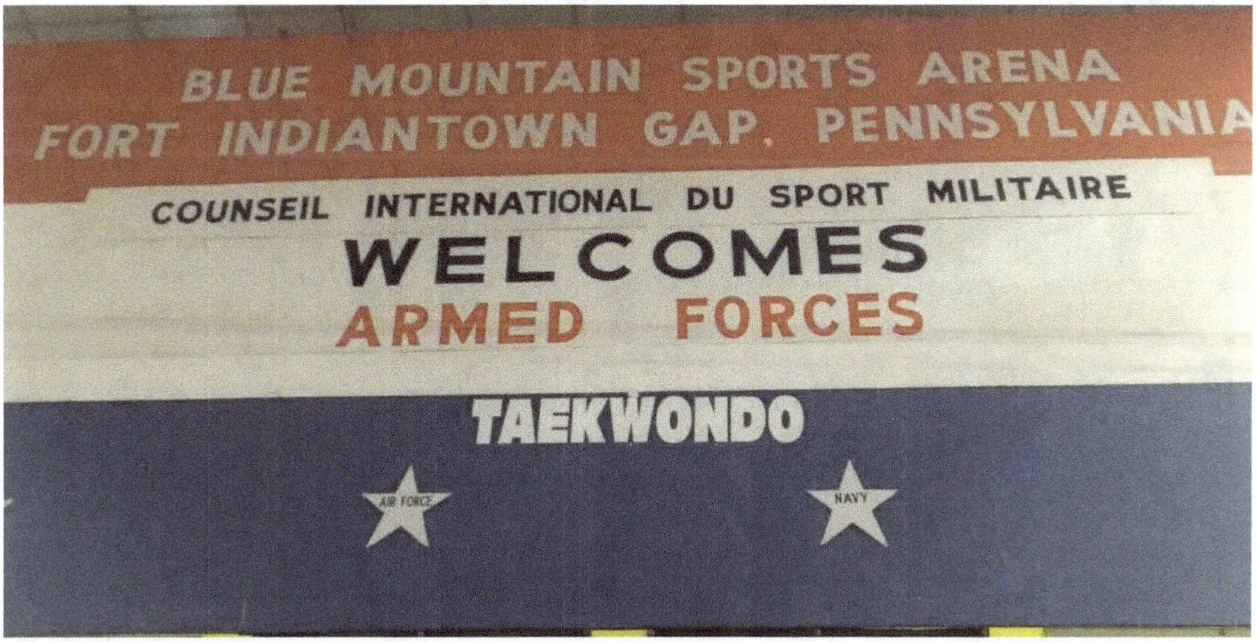
BLUE MOUNTAIN SPORTS ARENA
FORT INDIANTOWN GAP, PENNSYLVANIA
COUNSEIL INTERNATIONAL DU SPORT MILITAIRE
WELCOMES
ARMED FORCES
TAEKWONDO
AIR FORCE
NAVY

Mission Accomplished

# ABOUT THE AUTHORS

**MASTER LOUIS EDWARD DAVIS**

Louis Davis was born on October 11, 1970, in the city of Chicago, Illinois. In August of 1977 his family migrated to Minneapolis, Minnesota. By late 1985 he began training in martial arts at the age of 15 and by the summer of 1986 he began competing in local point Karate tournaments, some of which were hosted by Pat Worley.

During the 1993 USTU National Championships which were held in St. Paul, Minnesota, Louis came into contact with the All-Army Taekwondo Team, and after searching the venue he learned that there existed a team representing each branch of the US Armed Forces. After vowing to himself to one day become a member of one of these teams, he enlisted the following year.

In 1995 while at his first duty station, Fort Hood, Texas, he was approached by Sergeant Michael Bennett, who was a member of the All-Army Taekwondo Team competing in the heavyweight division. Under Sergeant Bennett and Sergeant Todd Angel's guidance, Louis went from an awkward novice to one of three team captains of the Fort Hood Taekwondo team.

In late April 1997, Louis along with fellow Fort Hood Taekwondo teammates John Swan, Nicolau Andradae, Ryan Lundy, Howard Clayton, and of course Sergeant Bennett and Sergeant Angel, attended the 1997 All-Army and Armed Forces Taekwondo trial and selection camp held at Fort Indiantown Gap, Pennsylvania. On April 27, Louis took the silver medal in the welterweight division during the Armed

Forces Championships, securing a place on the 1997 Army team.

In December of 1997, Louis relocated from Fort Hood, Texas to Harvey Barracks in Kitzingen, Germany. By October of 1998, Louis began to establish himself as a competitor throughout the Bavarian (Bayern) area, winning the Bavarian championships later that same year, holding said title until the year 2000. During this time Louis received guidance by phone from coach Rafael Medina on how to maximize his training and improve his skill set.

One of these many phone conversations led him to Georg Streif, the head coach of the German Armed Forces and German national team. Georg had invited him to attend one of his training camps which led to a major improvement in Louis' growing abilities.

In 1999, Louis returned to Fort Indiantown Gap and secured a place on the team after defeating Darryl Woods in the finals and defeating London Arevalo of the Air Force Taekwondo team during the Armed Forces Championships.

In 2000 Louis began visiting Pickens, South Carolina and entered a mentorship with former All-Army and national champion Reginald Perry. It was here that Louis began to research the history of the All-Army team.

The following year Louis returned to The Gap, securing a position as one of two middleweights on the Army team, and by January of 2003 relocated from Germany to South Korea, joining All-Army teammates Kevin Williams and Johnny Birch as a member of the 2nd Infantry Division's Taekwondo team.

During this four-year tenure in Korea as a member of the 2nd Infantry Division's Taekwondo team, Louis quickly established himself as a talented competitor, winning any competition held by Camp Casey or Camp Hovey respectively, winning both the 2003 and 2005 Friendship Cup (aka the Commander's Cup), the 2003-2005 Area 1 Championship, and the 8th Army championship.

In 2005 Louis took gold in the USA Taekwondo National Championships in San Jose, California, and his second national medal (silver) in Austin, Texas in 2009. In early to mid-2007, Louis was asked by Rafael Medina to attend the inaugural Taekwondo Hall of Fame banquet in Teaneck, New Jersey as a representative of the US Armed Forces, later becoming a technical advisor. Louis received recognition from the Hall of Fame for his efforts as a competitor. Louis continued to work as a technical advisor, representing the US Armed Forces until 2015.

After retiring from both the Army and Taekwondo competition, Louis returned to Minneapolis, Minnesota where he volunteers his knowledge and experience as a coach in support of a collective of Taekwondo school owners within the Twin Cities metro area, his former All-Army coaches and former teammates who operate Taekwondo schools themselves. He is currently a D Level referee, D level coach under USA Taekwondo, and a coach under the Amateur Athletic Union.

In addition to coaching, Louis is currently researching the history of the All-Army and Armed Forces Taekwondo Program by locating many of its pioneers, former champions, and team members.

**GRANDMASTER
MICHAEL RAY BENNETT**

My name is Michael Ray Bennett. I was born in Shreveport, Louisiana and attended Woodland High School.

I graduated from Southern University and A&M College with a BS in Computer Programming, from Houston Community College with an associate degree in Criminal Justice, and from Texas Southern University with a degree in Criminal Justice. I played sports in high school and college (football, basketball, and baseball).

I studied and trained in martial arts: Taekwondo, Ju Bushi Do, Aikido and Fu Kung. I am a Grandmaster in CMK and Sports Taekwondo, Master in KKW, Ju Bushi Do, Black Belt in Aikido and Black Sash in Fu Kung.

I proudly served in the Army for 24 years and am now retired. I was an All-Army Head Coach and athlete; I trained 50 All-Army athletes, producing 10 US national champions.
I am now retired as a federal police officer.

**GRANDMASTER RAFAEL MEDINA**

From February 15, 1973, to 1975 I started training Kyokushinkai with Sensei Miguel Acevedo and Sensei Fernando Caraballo. As a brown belt in Kyokushinkai he opened his first Karate dojo in Humacao, Puerto Rico, which was the first dojo ever in his town.

In 1977, I trained with GM Giovanni Rosario (RIP), International Taekwondo Federation (ITF-Young Brother Association). Around June or July 1978, Sensei Caraballo and Sensei Acevedo tested me for black belt in Kyokushinkai. In August I got married and in October I was into basic training (Fort Jackson, South Carolina).

I have accomplished many things in my life since I have been in the military. In 1985, I was among the first members of the newly formed Army Taekwondo Team that represented Fort Bragg in the 1984 and 1985 North Carolina and South Carolina state championships.

I established the motto "One team, one fight." Unifying the sport of taekwondo for all the armed forces. Now my motto is "One team, one fight, one family, stay strong."

In May 2019 I was recognized as a Grandmaster of Taekwondo by the government of Puerto Rico and certified by their House of Representatives.

In September 2019, the International Military Sports Council (CISM) selected me as the first military and the only person in the United States to represent the nation in the World Military Taekwondo Championships as an athlete, coach, international referee, and first Latino soldier among 140 member-nations of CISM.

After I retired from the service, I began working with children as a coach for the Liberty County Recreation Department's sport Taekwondo team. I advise kids to stay away from guns, no violence at home and school, no drugs, no bullying, and to listen to their teachers and parents. Kids who train in Taekwondo not only become combat capable, but also become better human beings by learning discipline and respect.

But my greatest accomplishment came as a complete surprise when I received the news that I was being nominated for the Taekwondo Hall of Fame as The Outstanding Pioneer Armed Forces Player Award.

I also was worthy of recognition twice by the U.S. Taekwondo Champions as the Pioneer Award of the year for my performance in the role of an athlete, coach, referee, first to organize the Armed Forces Taekwondo reunion, and tournament coordinator.

I had produced more qualified athletes for the Armed Forces Taekwondo teams, many of which continued forward to become All-Army, Armed Forces, CISM and US National medalists, coach and even president for the CISM Taekwondo committee. I will continue to guide future military athletes to the Army Taekwondo Team as well as civilians to continue with the discipline and tradition of the martial arts.

Around 1986, The Department of Sport/MWR for the first time paid us all competition expenses. We were literally the first soldiers (Pedro Laboy, Mark Green, Leo Oledan, and Rafael Medina) representing the Army and Armed Forces in a Taekwondo competition.

SFC Bennett and I opened a Taekwondo dojang on base (Fort Bliss). During this time, I was getting SFC Bennett ready for the All-Army Taekwondo team trial. In 1995, Coach Bobby Clayton asked me if I would like to be his assistant coach.

My achievements have been thanks to the support of the people around me, because without them I would never have gotten good results. They are the ones to be "blamed" for me being where I am today: Luis Diaz, Miguel Acevedo, Fernando Caraballo, Giovanni Rosario, Pedro Laboy, Bruce Harris, Paul Boltz, Michael Bennett, Bobby Clayton, Bongseok Kim, and the Puerto Rico Taekwondo Federation.

# ★★★ VETERAN'S CREED ★★★

## I'M A VETERAN

I HAVE SEEN AND DONE THINGS THAT MANY WILL NOT UNDERSTAND.

I'M A WARRIOR AND

MEMBER OF A TEAM SPANNING THE WORLD.

I HAVE SERVED MY COUNTRY PROUDLY & NOW STAND BY TO SERVE MY

BROTHERS AND SISTERS IN ARMS.

**I WILL NEVER ACCEPT DEFEAT I WILL NEVER QUIT**

I WILL NEVER LEAVE A FALLEN BROTHER OR SISTER. IF THEY ARE HURT, I

WILL CARRY THEM.

IF I CAN'T CARRY THEM, I WILL DRAG THEM. I WILL HELP THEM FACE THEIR

ENEMIES. TO INCLUDE THE DEMONS FROM WITHIN.

THEY ARE MY BROTHERS AND THEY ARE MY SISTERS.

## I AM A VETERAN!

www.ingramcontent.com/pod-product-compliance
Lightning Source LLC
Chambersburg PA
CBHW082037300426
44117CB00015B/2514